The GSM Network

The GSM Network

GPRS Evolution: One Step Towards UMTS

Second Edition

Joachim Tisal
Siris, France

Translated by John C. C. Nelson
School of Electronic and Electrical Engineering, University of Leeds, UK

JOHN WILEY & SONS, LTD
Chichester · New York · Weinheim · Brisbane · Singapore · Toronto

Baffins Lane, Chichester,
West Sussex, PO19 1UD, England

National 01243 779777
International (+44) 1243 779777

e-mail (for orders and customer service enquiries): cs-books@wiley.co.uk
Visit our Home Page on http://www.wiley.co.uk
or http://www.wiley.com

Other Wiley Editorial Offices

John Wiley & Sons, Inc., 605 Third Avenue,
New York, NY 10158-0012, USA

WILEY-VCH Verlag GmbH
Pappelallee 3, D-69469 Weinheim, Germany

John Wiley & Sons Australia, Ltd, 33 Park Road, Milton,
Queensland 4064, Australia

John Wiley & Sons (Canada) Ltd, 22 Worcester Road,
Rexdale, Ontario, M9W 1L1, Canada

John Wiley & Sons (Asia) Pte Ltd, 2 Clementi Loop #02-01,
Jin Xing Distripark, Singapore 129809

Library of Congress Cataloging-in-Publication Data
Tisal, Joachim.
The GSM Network
GPRS evolution : one step towards UMTS / J. Tisal ; translated by John C.C. Nelson.
p. cm.
Includes bibliographical references and index.
ISBN 0-471-49816-5 (alk. paper)
1. Wireless communication systems. 2. Mobile communication systems. I. Title.

TK5103.2 .T57 2001
621.382--dc21 00-068515

British Library Cataloguing in Publication Data
A catalogue record for this book is available from the British Library

ISBN 0 471 49816 5

Typeset by Deerpark Publishing Services Ltd, Shannon, Ireland.
Printed and bound in Great Britain by Antony Rowe, Chippenham, Wiltshire.
This book is printed on acid-free paper responsibly manufactured from sustainable forestry, in which at least two trees are planted for each one used for paper production.

Contents

Preface

This work presents the concepts which inspire the technology of the cellular radio telephone, the hardware architecture of a cellular network, the components of the network and the foreseeable developments of networks of this type.

GSM is a certain success and numerous indications show that users want data services on mobile networks in the short term. Nevertheless, GSM data services suffer from the following constraints: a data rate limited to 14.4 kbit/s, the use of circuit switching and lack of a direct interface to data networks.

High speed data services (High Speed Circuit Switched Data (HSCSD)), defined in the GSM standard for Phase 2+, offer higher rates. To achieve these rates, several circuits are used in parallel. This approach to Internet access is inefficient, since radio channels are reserved for a single communication although the exchanges of data are sporadic.

For GSM operators, the use of circuit switching for data transport imposes a large penalty on the management of radio resources. Operators are required to use packet switching at the radio interface. Packet switching permits optimisation of the use of data channels at the radio interface and charging of users by volume instead of duration. To respond to this need, the European Telecommunications Standards Institute (ETSI) has defined a packet mode data service: the General Packet Radio Service (GPRS).

Mobile terminals see GPRS as a level 2 access layer to data networks (X.25 or IP). GPRS is a path which permits convenient connection of an application located in a mobile terminal to data bases; because of the high data rate characteristics (115 kbit/s), low delay and charging by usage, a user can remain connected for a long time without consuming radio interface resource. Target applications for GPRS are the mobile office, Internet and intranet connection, telemetry, radio guidance and security.

GPRS can be introduced into a network without modification of the radio subsystem. GPRS uses the same modulation, the same frequency bands and the same frame structure. The physical layer regards the new packet mode data channels as voice service traffic channels. In contrast, new mobile terminals will appear which can serve as a radio modem connected to a portable computer or a Personal Digital Assistant (PDA).

The evolution of GSM will accentuate the convergence of mobile and fixed usage and quality (*a portable telephone with Internet capacity*). Operators will be able to offer levels of service quality by means of different user profile service definitions with bandwidths on demand.

GPRS is an evolution of GSM networks which prepares for the introduction of third generation networks. The major differences are the data rate, which can reach 2 Mbit/s with UMTS (Universal Mobile Telecommunications System), and the nature of the terminals which will be multimode.

Introduction

1

1.1 Historical Perspective

From ancient times right up to the end of the 19th century, writing was the only means of communication between two people separated by a large distance. In this period prior to telecommunications, information was conveyed at a speed limited by the message carriers. From the beginning of the 20th century, telecommunications has bridged the distances between people who wish to communicate. We shall briefly run through the major stages in the history of telecommunications leading up to the birth of Global System for Mobile Communications (GSM), before tackling GSM in detail in terms of its protocols and network equipment, and the characteristics of network terminals.

Progress in the theory and practice of telecommunications was made in the 19th century.

In 1876, the Canadian scientist Alexander Graham Bell (1847–1922) invented the telephone, the first means of modern telecommunication. Communication was demonstrated between two fixed instruments linked by a pair of wires (the telephone wire-pair).

In 1887, the German physicist Heinrich Hertz (1857–1894) discovered 'hertzian waves', now known as radio waves.

In 1896 in Bologna, the Italian physicist Guglielmo Marconi (1874–1937) succeeded in achieving the first radio transmission. The experiment took place in his attic, where he had set up his laboratory. Marconi succeeded in remotely controlling an electric bell by radio, the distance between the transmitter and the receiver being a few metres. Marconi patented his invention, which he called 'wireless telegraphy'. In order to perfect his invention, Marconi set himself up in England between 1897 and 1901, where he created the first transatlantic radio transmission between Cornwall and Newfoundland in 1901. The radio link opened up the era of long-distance telecommunications. In less than 10 years, Marconi had developed a transmission technique that was both efficient and reliable, and in 1909 he received the Nobel Prize for Physics in recognition of his work on 'hertzian waves'.

Hertzian or electromagnetic waves allow communication between two fixed points, but furthermore, they offer an ideal solution to the problem of communicating with moving vehicles of all types; boats, planes, satellites, cars or pedestrians, whatever the distance between the parties.

From the beginning of the 20th century, police forces in Europe and North America have been equipped to communicate with patrol vehicles.

At the beginning of the 1950s, the Bell Telephone company in the United States introduced a radio telephone service to its customers. For the first time, the radio telephone, until then reserved for institutions, became commonly available to the public at large. However, the radio telephone network was only able to accommodate a small number of subscribers.

To counter the increasing demand for connection to the service, it was necessary to invent new techniques for sharing the frequency bands between the largest number of subscribers, and to improve the operation of the network. The lack of radio spectrum available for radio telephony slowed down the expansion of this form of communication.

In 1964, the concept of shared resources was introduced into radio telephone networks. Networks with this facility dynamically allocate a radio channel to a new call for its duration. The system selects one from the range of frequencies that are free, and allots it to a new call. This development was important because the management of frequencies, which had been fixed until that time, was now dynamic. The consequence of this was that the network could now have more subscribers on it than there were radio channels.

Radio telephone frequencies were a scarce resource and there was a constant need to optimise their utilisation. In 1971, the Bell Telephone company in the United States proposed a new cellular network concept in response to the demand for a mass market radio telephone system, using a limited frequency band, allocated by the Federal Communications Commission (FCC). The Bell Telephone Company introduced the 'Advanced Mobile Phone Service' (AMPS), putting into effect the cellular concept. The system was first trialled in Chicago, and has been operational there since 1978.

The cellular network principle is based on the sub-division of the geographical area covered by the network into a number of smaller areas, called cells. In each of the cells, a fixed station acts as a transmitter-receiver serving all the mobile stations situated within the cell boundary. The advance that this represented with respect to the previous arrangement was the increased number of mobile stations that it allowed and the reduction of emitted radio power. Today, each base station indeed covers a smaller area, but the corollary is the need for management of the changeover of cells when a mobile station moves from one cell to a neighbouring one. A base station controls the group of frequencies allocated by the network to that cell, and the group of subscribers that are present in the cell. When a subscriber wishes to initiate a call, the base station allocates a transmitter frequency to him. As soon as the subscriber moves into another cell, he comes under the control of another base station, which allots him a frequency different from the first. This liberates the frequency used in the original cell. The novel features brought by the cellular concept are dynamic changing of transmission frequency for a subscriber's terminal during a call as a function of the movements of the mobile, and the re-use of frequencies in cells separated far enough apart. These features allow a larger number of simultaneous calls to take place in the network, and therefore make possible a larger number of subscribers.

In 1982, the FCC in the United States standardised the AMPS system specifications, and this became the single radio telephony standard for North America.

Several cellular radio networks entered service around the world in the 1980s. These were analogue systems and were designed to mutually incompatible standards, with each country exercising its right to select its standard.

Today, there are many cellular radio networks, and radio telephony has become commonplace. It is now possible to make calls anywhere, and at any time, and to transmit and receive both speech and data. Multimedia telecommunications for the public has become a reality, transporting equally well sound, text, and pictures. Data is a mixture of these different types of information.

A variety of services are offered to the public:

- **One-way radio paging**: the transmission of alphanumerical messages from a distribution centre to a subscriber within a defined geographical area, without the facility of acknowledgement or reply (Eurosignal, Alphapage™).

- **Satellite-based systems**: these systems transmit data and have world-wide coverage.

- **Radio telephony**: this uses either analogue or digital transmission. It offers private circuit and shared-resource radio and transports speech, messages or data.

- **Cordless telephony**: this can be operated just on customers' premises, or can be a public network, e.g. Telepoint systems in the UK (now no longer commercially available), and 'BiBop' in France.

- **Telepayment**: popular applications of this facility are motorway tolls, parking and control of admission.

- **Local cordless networks**: these networks give mobility to a group of terminals within a restricted local network. Common current applications are portable terminals for use in warehouses and shops, for monitoring stocks and updating data bases in real time.

In Europe, each country has gone its own way in adopting a standard for analogue radio telephony. The UK and Italy have chosen the American standard, under the name 'Total Access Cellular System' (TACS), whereas the Scandinavian and Benelux countries, and the French operator Société Francaise du Radiotéléphone (SFR) have selected the Nordic Mobile Telephone standard (NMT). France Telecom has developed the Radiocom 2000TM standard, and Germany has preferred the C-Net standard. All of these standards use analogue transmission, have small capacity in terms of subscribers and traffic carrying, and they require a large number of frequencies. In all of the world's major cities (Los Angeles, New York, Chicago, Paris, London, and Rome) the radio telephone is the victim of its own success. Network operators can no longer meet the demand from subscribers, which is outstripping network capacities. This situation is caused by the lack of a new radio spectrum for the networks, and the limitations of the technology deployed. It has been necessary to develop two techniques in order to counter these problems; time division multiplexing, and changing from analogue to digital transmission. Table 1.1 lists the main analogue standards in use in Europe.

Due to the increased mobility of radio telephone users in the European region, the problems caused by the incompatibility of analogue standards in Europe and the technical development of the network imposed by its saturation, European Telcos have been driven to co-ordinate the development of a common solution. The aims of this solution are firstly to offer users a single means of accessing radio telephony service, and secondly to provide a large-volume market to manufacturers, who thus need only support a relatively small number of product variants. They can thus realise economies of scale, and can offer lower equipment prices.

Table 1.1: Analogue standards in Europe

Country	Standard
Germany	C-450
Belgium	Mobilophone 2, NMT
Denmark	NMT
Spain	NMT
Finland	NMT
France	Radiocom 2000, NMT
Italy	TACS
Norway	NMT
Netherlands	NMT
United Kingdom	TACS
Sweden	NMT

1.2 Radio Transmission Fundamentals

1.2.1 Modulation

Figure 1.1 shows two signals $F(t)$ and $F'(t)$. They have the same amplitude and frequency, but $F'(t)$ has a phase difference of ϕ with respect to $F(t)$.

Each of these three parameters (amplitude, frequency and phase) can carry information, and are the basis of three types of signal modulation:

- amplitude modulation;
- frequency modulation;
- phase modulation.

Amplitude modulation
The amplitude of the carrier wave is proportional to the magnitude to be transmitted. The main advantage of this modulation technique is the simplicity of its implementation, and its main drawback is its sensitivity to radio

interference. This modulation format is widely used in AM radio broadcasting and for television.

Frequency modulation

The instantaneous frequency is varied linearly with the signal. This modulation format is less sensitive to interference than amplitude modulation but requires a larger bandwidth. In order to transmit a signal having a frequency deviation L, the required frequency bandwidth is $B = 2 \times L$. This modulation format is commonly used for FM broadcasting.

Phase modulation

The phase of the carrier signal is varied linearly with the signal strength with respect to a reference signal, but the phase excursion is limited to 2π. The bandwidth required is the same as for frequency modulation.

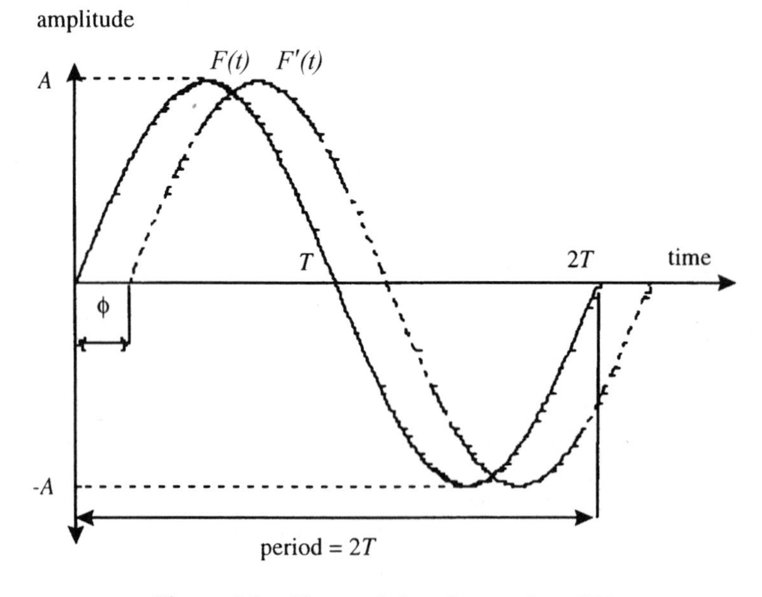

Figure 1.1: Characteristics of a waveform $F(t)$

1.2.2 Transmission

There are two fundamental modes of transmitting information in telecommunications. These are:

- analogue;
- digital.

Analogue transmission

Figure 1.2 shows the main elements of an analogue radio transmission system for telephony. In analogue transmission, the physical quantity carrying the information (amplitude, frequency, and phase) varies in a direct relationship to the information to be transmitted. In this mode, the principle of proportional variation operates.

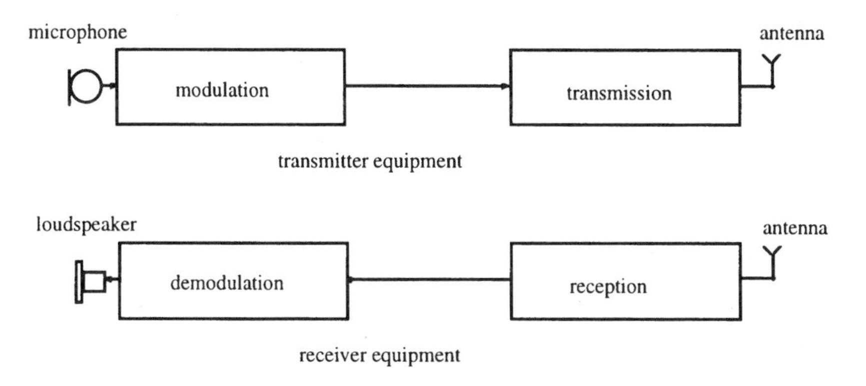

Figure 1.2: Analogue radio transmission system

In the transmitting equipment, the microphone fulfils two roles. It serves as both an acoustic sensor and a transducer, for converting the sound signal into an electrical form. In other words, it receives a sound wave and transforms it into an electric signal whose voltage and frequency vary in sympathy with the amplitude and frequency of the acoustic signal that is picked up. In the modulation unit, the carrier wave is modulated by the signal output of the microphone according to the modulation technique utilised (amplitude, frequency or phase). The processing of the signal that is required takes place in the transmitter equipment, and it is transmitted from the radio antenna. In the receiver, the signal undergoes the reverse process, finally reconstituting, via the loudspeaker, the sound signal that was received by the microphone. The receiving antenna picks up the incoming radio wave, which is then amplified before being sent to the demodulator, which extracts the underlying signal from the carrier before feeding it to the loudspeaker or earpiece. The loudspeaker is (often) an electromechanical transducer, which converts electrical energy into mechanical energy in the form of sound waves. These processes are not described further here.

Digital transmission

This transmission mode requires more complicated processing than analogue transmission. Before transmission, the signal must be converted into a digital format. This occurs in three operations for a signal $s(t)$. These are:

- sampling;
- quantization;
- coding.

Sampling of the signal consists of slicing the continuous signal $s(t)$ at regular time intervals into discrete values, the magnitudes of which are equal to the amplitude of the signal at a given instant t.

The quantization process assigns values to the measured amplitudes according to a conversion law. The choice of conversion law depends mainly on the range of measured amplitude variation, but it ensures that either the quantization values are proportional to the signal or that there is a dynamic relationship between neighbouring samples. In general, quantization follows a logarithmic law, which guarantees almost uniform precision, even for amplitudes falling within a narrow range.

Coding is the final operation in signal digitisation. A digital value is assigned to a sample as a function of the number of digits available. As the information signal varies, a representative physical quantity is turned into a digital value. To achieve this, the range of magnitude of the signal $s(t)$ is divided into n intervals, for example into 128 positive levels and 127 negative levels, giving a range of variation covered by 255 values. Eight bits, that is one byte, are necessary to give a unique code to each of these levels, and it is these 8-bit coded samples which are the information that is transported down the transmission channel.

The advantage of digital transmission over analogue transmission lies in the ability to add redundant information to the true information. This redundancy provides a means of detecting errors incurred by the signal during its transmission, and offers various means of correcting them. This is impossible with analogue transmission.

With digital transmission, a binary representation of the signal is what is transported, rather than the variations of amplitude $s(t)$ (for example, in frequency modulation, a binary digit '0' could be represented by the frequency F_1 and the binary digit '1' by the frequency F_2). The transmitted codes obey a defined format of N bits known to the transmitter and receiver; the bits are transmitted in a defined order, for example the least significant first. Thus, for example, the number 1894 can be distinguished from 4981 although they are composed of the same digits. The format and bit ordering are part of the transmission protocol existing between the transmitter and receiver. The receiving equipment controls this format, is able to detect violations, and is able to act to correct the errors according to the characteristics of the transmission code utilised. Figure 1.3 shows a schematic representation of a digital transmission system.

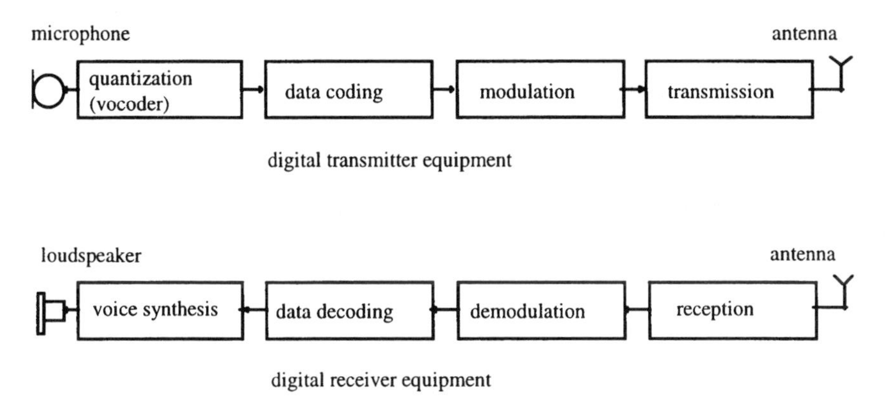

Figure 1.3: Digital transmitter and receiver systems

The microphone transforms the acoustic signal into an electrical signal, which is digitised by a vocoder or COder-DECoder (Codec). The acoustic signal input is transformed by the Codec into a series of samples, the amplitudes of which are coded. In the coder, the data stream is ordered, coded and output as a serial bit-stream before modulating the transmitted carrier wave according a digital modulation technique. At the receiver, the bit-stream undergoes exactly the reverse of the operation at the transmitter. After extraction of the information from the radio signal, the sound wave is reconstructed before being fed to the earpiece or loudspeaker.

Table 1.2 shows the stages in the change of format of the information in a digital transmission system.

Note: A telephone signal is considered to be an analogue signal occupying a channel bandwidth of 0–4 kHz. Signal theory sets a rule to be obeyed in sampling in order not to lose information. A signal can be reconstructed without loss of information from a series of samples taken at a rate equal to at least twice the bandwidth of the signal, i.e. 8 kHz for a telephone signal. A more detailed description of these processes is beyond the scope of this hook. The interested reader might refer to the publications of B Picibono [58] and J C Radix [60].

The choice between modulation formats (amplitude, frequency or phase) and between transmission modes (analogue or digital) depends on system parameters such as the transmission environment, the interference characteristics, and the information rate to be transmitted.

Table 1.2: Transformation of information as it passes over a digital transmission system

Physical	Logical
Sound signal	
	Digitisation
	Coding
Modulation	
Transmission	
Reception	
Demodulation	
	Decoding
Synthesis	
Sound signal	

1.3 Impairments to Radio Transmission

In transmission by either copper wires or optical fibres, where the medium is well defined and controlled, the impairments can be theoretically modelled. This is more problematical with radio transmission. The impairments are known, but because it is impossible to completely control the transmission environment, their effect as a function of time is unpredictable, and hence it is difficult to model radio transmission dynamically. We shall now examine the main sources of radio interference.

The effect of the ground and obstacles

Radio waves are reflected by not only the ground, but also by trees, buildings, etc. These reflected waves arrive out of phase with the direct transmitter-to-receiver beam, and the receiver detects the sum of the direct and reflected waves. The reflecting surfaces between the transmitter and receiver cause interference between the direct and reflected beam, and this gives rise to an impairment known as fading. Figure 1.4 illustrates this situation.

Co-channel interference

Distant radio transmitters transmitting on the same frequency as that used by a particular radio link will disturb reception, even if they are very far away. This creates an impairment known as co-channel interference.

Intermodulation distortion

Two radio transmitters transmitting on different frequencies can disturb each other when they are too close or are too powerful. This impairment is known as intermodulation.

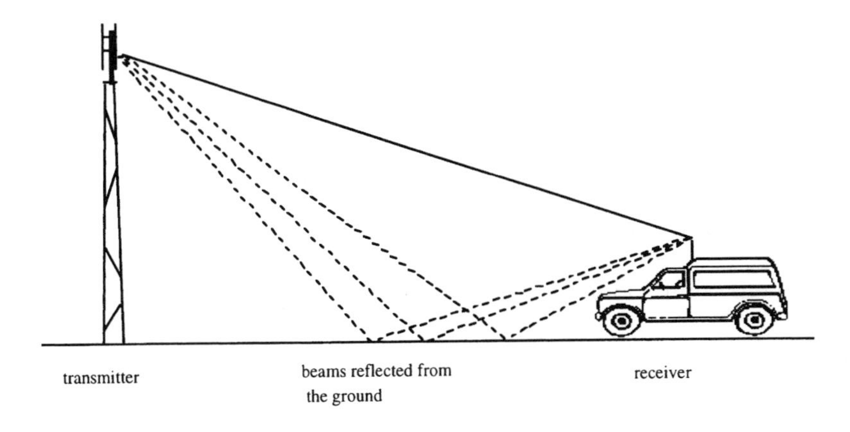

Figure 1.4: Reflection of radio waves from the ground

Background noise

Background noise arises from the sensitivity of the receiver. The receiving antenna picks up thermal noise generated by the receiver electronics at the same time as the signal, and these are both amplified together. Filtering allows an improvement in the signal-to-noise ratio.

Atmospheric noise

Radio waves created during storms and other atmospheric phenomena can interfere with radio signals. They are characterised by very short intense noise bursts. Atmospheric noise is important in the low frequency bands.

Industrial noise

All electrical equipment can be the source of radio 'pollution'. Electrical motor and dynamo brushes and the ignition systems of combustion engines are also sources of parasitic interference.

1.4 Creation of the GSM Group

In 1982, the Conference of European Posts and Telecommunications administrations (CEPT) created the Groupe Special Mobile (GSM) (now known more widely as Global System for Mobile communications). CEPT charged the GSM group with developing European standards for mobile radio-telecommunications in the frequency band reserved for this purpose in 1978 by the World Administrative Radio Conference (WARC).

In 1982 CEPT defined the following bands:

- the 890–915 MHz band for transmission from mobile stations;
- the 935–960 MHz band for transmission from fixed stations.

The aims identified by CEPT for the GSM standards were:

- a large subscriber base;
- world-wide compatibility;
- efficient use of radio spectrum;
- wide availability;
- adaptability to traffic conditions;
- a quality of service comparable to that of the fixed network;
- an affordable cost to subscribers;
- the ability to access the network from either mobile or portable handsets;
- basic telephone service with special services such as control of closed user groups.

In 1987, the GSM group settled on the main technical options for the mobile radio telephony standards. These were the following:

- digital transmission;
- time-division multiplexing of radio channels;
- encryption of radio channel transmission;
- a new compression algorithm for reduced data rate compared to the coding laws used in telecommunications (the μ-law in Europe, and A-law in North America.

1.5 Digital Standards

The advantages of digital transmission compared with analogue transmission for radio telephony are:

- a better telephony quality;

- an increased potential number of subscribers served by the network;
- the possibility of new services (data, text, facsimile, and image transmission);
- confidentiality through radio transmission encryption;
- the ability to control transmitter power;
- the ability to offer access to the ISDN network.

Table 1.3 lists the current principal radio telephony standards.

Table 1.3: Digital standards for radio telephony

Standard	Application/country	Cell size
CT-2	Cordless telephony/Europe	200 m
GSM	Radio telephony/Europe	200 m to 35 km
1S54	Radio telephony/North America Japan	200 m to 35 km
1S95	Radio telephony/North America	200 m to 35 km

Spectral efficiency of digital techniques

Table 1.4 lists the spectral efficiency of different transmission standards, measured in bits/s/Hz.

Table 1.4: Spectral efficiency in the digital standards

Standards	Spectral efficiency (bits/s/Hz)
CT2	0.72
GSM	1.35
1S54	1.62

In the USA, the Telecommunication Industry Association has published two provisional standards: IS-54 in 1990, using Time Division Multiple Access (TDMA), and IS-95 in 1993, using Code Division Multiple Access (CDMA). These digital standards are compatible with the Advanced Mobile Phone Service (AMPS) and TDMA and offer a migration path between the former analogue system and the new digital system. The new standards allow the number of subscribers on the network to be trebled.

Note: TDMA is a temporal multiplexing technique. Time division multiplexing consists of dividing the time into equal intervals called frames. A frame is composed of elementary time intervals, or timeslots, (see Figure 1.5); each timeslot constitutes a data transmission channel in which blocks of coded information are transmitted.

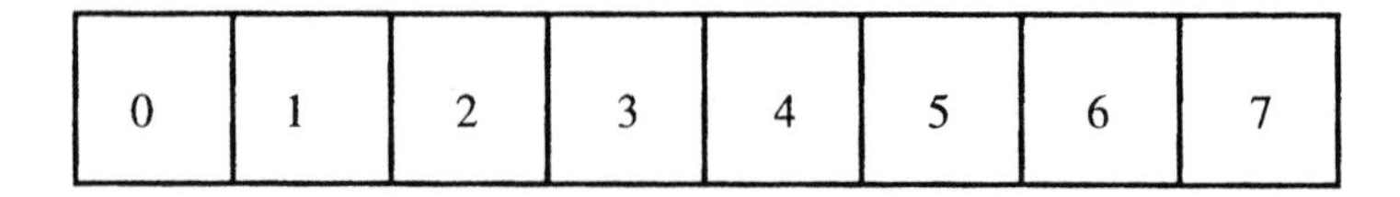

Figure 1.5: The structure of a TDMA frame

In September 1987, the network operators of 13 European countries (Germany, Belgium, Denmark, Spain, Finland, France, Ireland, Italy, Norway, Netherlands, Portugal, UK, and Sweden) signed a Memorandum of Understanding (MOU) for the concerted implementation of a European digital cellular network. Other countries were also later to join this agreement. There has since been an enlargement of the group of signatories to include non-European countries. The GSM standard has also now been adopted by countries in Africa, Asia and Australia.

In 1989 the work of the original GSM group was transferred to the SMG committee of the European Telecommunications Standards Institute (ETSI), who are continuing the standardisation tasks.

Five sub-groups make up the SMG. These define:

- services offered;
- the radio interface;
- network aspects;
- data transmission;
- the Subscriber Identity Module (SIM).

The SIM is a memory card that holds data describing the subscription details.

1.6 History of Speech Coding

The International Telecommunications Union (ITU-T) (formerly CCITT) has standardised speech coding algorithms over a long period. The first digital speech coding algorithm was Pulse Code Modulation (PCM), which became ITU recommendation G.711 in 1972. However, the basic bit rate of 64 kbit/s was much higher than common data transmission rates, and several hundred times higher than that of the linguistic information content of speech. There then ensued a race for bit-rate reducing coding algorithms, with the requirement of maintaining tonal speech quality. A speech quality scale is now available – the Mean Opinion Score (MOS). MOS is a measure of speech quality in a telephone network and has five categories (5: excellent, 4: good, 3: fair, 2: poor, 1: unsatisfactory).

With the first PCM method, each speech sample is independent of the previous one. In order to attain the goal of reducing bit rate whilst retaining tonal quality, one way is to derive the value of a speech sample from the previous one.

Speech synthesis systems such as Linear Predictive Coding (LPC) predict the current sample from a linear combination of past samples using a least mean squares criterion. But though they achieve high efficiency, they have poor tone quality. The Adaptive Differential PCM technique (ADPCM) offers an alternative method of predicting a speech waveform from past samples. The algorithm operates at 32 kbit/s and became the G.721 standard in 1983.

The Low Delay Code Excited Linear Predictive algorithm (LD-CELP), which is the G.728 standard, has a bit rate of 16 kbit/s. In this technique, speech synthesis is carried out using a five-sample unit and the parameters for the synthesis are refreshed every 20 samples. Backward processing, in which linear prediction coefficients are obtained from stored past samples, enables low delay to be achieved and dispenses with the need to transmit LPC coefficients.

The CS-ACELP is the candidate for the 8 kbit/s speech coding algorithm standard, initiated by the SG15/ITU-T speech coding expert from Brazil. The performance requirements are:

- A quality equal to the 32 kbit/s ADPCM in error free conditions;
- An MOS reduction of less than 0.5 compared to error free ADPCM;
- An encoding-decoding delay time less than 10 ms.

The difficulty of meeting the delay requirement prevented this scheme from being adopted. Therefore, in 1992 the delay requirement was relaxed from 10 ms to 32 ms on the assumption that echo-cancellers would be become standard equipment. In 1992, the Nippon Telecommunications and Telegraph company (NTT) proposed the Conjugate Structure Vector Quantiser Code

Excited Linear Predictive algorithm (CS-CELP) and showed that this algorithm came close to meeting the requirements. Later in the same year, France Telecom and the University of Scherbrooke jointly proposed the ACELP algorithm. Several organisations tested these two candidate algorithms and found that they both met the requirements, except the quality requirement under conditions of background noise. The two teams agreed to form a development group to build a better algorithm and some time later AT&T joined the group. In 1994, the development group proposed a new algorithm – CS-ACELP. The quality tests demonstrated that this algorithm meets all the requirements and it was approved as the ITU-T recommendation G.729 in February 1995.

Cellular Concepts

2

2.1 Frequency Re-use

This chapter sets out the constraints arising from three network aspects; subscriber mobility within the network, the reach of the network, and the characteristics of radio propagation. It then describes the solutions to these problems adopted in cellular networks.

Limitations imposed by subscriber mobility have an impact on the characteristics of subscriber terminals. The subscriber terminal must be intrinsically mobile in respect of its use in a vehicle (e.g. boat, car or motorbike), or as a handheld instrument. This means that the terminal must be of restricted size and weight so that it is easy to carry and manoeuvre. However, a compact and lightweight unit implies a power supply of limited capacity, which is an important factor if a reasonable duration of stand-alone operation is desired. Other important aspects of mobility concern the network, including the need to locate subscribers, the ability to follow their movements, and the capacity to monitor their accessibility (a subscriber who turns off his handset disappears from the network). A final important characteristic of mobile terminals is their performance in noisy environments – city streets are rarely quiet places.

The radio beam is the link between the subscriber's mobile terminal and the fixed infrastructure of the network. The fixed network terminals containing transmitters and receivers are known as 'base stations'. Since the output power of the subscriber terminal is limited by its battery supply capacity, the emitted radio power is weak. Moreover, it should be borne in mind that the attenuation of a radio signal is proportional to the fourth power of the distance between the transmitter and receiver. That is, the emitted power P_c, of a transmitter and the power, P_r, arriving at the receiver are related to the distance, d, by the equation $P_r = P_c/d^4$. Therefore, the greater the distance, d, the weaker the received signal. For example, if we consider three receivers R_1, R_2, and R_3 situated at distances d_1, d_2, and d_3 from the transmitter, where d_2 is double d_1 ($d = 2 \times d_1$), then the signal power P_{r2} arriving at receiver R_2 will be 16 times weaker than that received (P_{r2}) by receiver R_1. If the distance d_3 is three times d_1, the received power, P_{r3}, will be 81 times weaker than P_{r1}.

This property shows that if the emitted power is limited, the range of the signal will also be limited. For a significant coverage area, with subscriber terminals having low transmitter power, it is therefore necessary to have a substantial number of base stations. The corollary of this is that, given that the weak transmitted power limits the coverage area, it is possible to re-use the same frequency in zones separated sufficiently far apart without suffering from co-channel interference.

In radio telephony, the zone covered by a base station is known as a 'cell'; hence the title of 'cellular' radio telephony for networks engineered for the best performance within the given limitations. Frequency re-use is an important concept in public radio telephony because the frequency spectrum is a scarce resource and this type of 're-cycling' optimises its usage.

In cellular radio, there is a direct relationship between the number of available frequencies and the number of calls possible at any given time. The number of frequencies available determines the traffic capacity of the network. In a given zone, the smaller the cell size, the greater is the number of simultaneously usable frequencies. This fact is exploited to the full in urban areas where the traffic density is high.

In a network, the cell is the basic element used in the architecture of a zone which uses a group of frequencies. To use any one of these frequencies in two different cells C1 and C2, it must be arranged that they are at least two cell diameters apart in order to limit co-channel interference. Figure 2.1 shows a cell architecture, with the cells represented as hexagons.

In a GSM network, a cell pattern made up of nine cells is defined with which to structure the geographical area for cellular telephony. This is shown in Figure 2.2. In general, the base stations are located at the intersection of three cells in order to optimise their number. Figure 2.2 shows the ideal use in which a base station situated at the intersection of cells A1, A2 and A3 can serve the three cells.

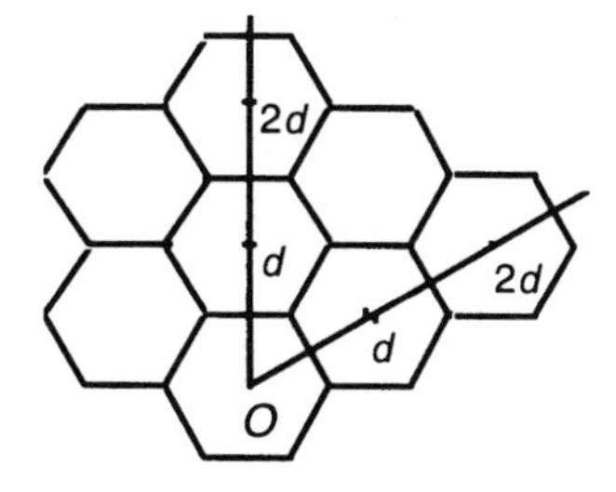

Figure 2.1: The basis for the re-utilisation of frequencies

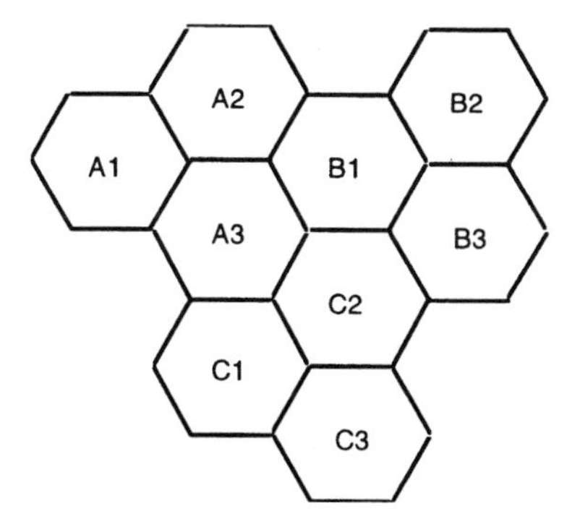

Figure 2.2: A pattern of nine cells

A cluster is the logical grouping unit of one or more cells between which bearer handover takes place.

2.2 Mobility

The freedom of users to move imposes the following requirements:

- locating subscribers in the network;
- monitoring subscriber movement in the network;
- maintaining the link between the network and the subscriber while the user changes cell.

When the network perceives that the subscriber who is conducting a call will cross the boundary between two cells, it transfers the call between the two cells in a way that disrupts the call as little as possible. This intercellular transfer, known as handover, is technically complex.

Handover is the process of switching a call in progress from one physical channel to another. Two forms of handover exist – intra-cell handover and inter-cell handover. Intra-cell handover is the transfer of a call in progress on a physical channel of one cell to another physical channel of the same cell. Inter-cell handover involves transferring the call between two cells.

In analogue cellular networks, when the mobile terminal changes cell, communication is first interrupted in the active cell, before being re-established in the new cell. In a GSM network, the mobile terminal is equipped with several demodulators, and it can thus stay in contact with two base stations simultaneously, in such a way as to avoid breaking the link to the fixed network, and this is the first main advantage. The second advantage follows from the first in that the ping-pong effect, resulting from a mobile terminal moving in an area along the boundary between two cells, is avoided.

Roaming is the movement of the mobile terminal from one part of the network coverage area to another part, whilst retaining the network capability to make or receive calls in both areas.

The *attach* process is the way in which a mobile station within the network coverage area in which it has rights notifies the network that it is operative. The reverse process is *detach*, in which the mobile terminal reports that it is inoperative.

2.3 Multiple Access Design

The choice of multiple access model is an important feature of the design of a cellular mobile radio system. A solution can be found by combining the basic multiple access methods: Frequency Division Multiple Access (FDMA), Time Division Multiple Access (TDMA), and Code Division Multiple Access (CDMA). Practical multiple access schemes are usually hybrids of these methods. The way in which the system resolves the problem of ordering and separating the multiple radiated signals at the receivers influences the total system design. Determining viable multiple access designs is currently receiving much research and development effort around the world. The large range of different combinations of techniques means that there are significant technical challenges, and it is unlikely that a unique optimum solution exists.

In a single-cell scenario with ideal conditions, all properly designed multiple access models, including hybrid schemes, would be equivalent in their capacity to separate user signals at the receivers. Models perform differently when the frequency dependence and delay variability of real multi-path radio channels come into play. Comparing the three basic multiple access methods, FDMA suffers relatively little due to the sporadic nature of user signals because this property of the signals is not affected by the frequency dependence of the channels. Signals are moderately impaired in TDMA systems because the signal burstiness is affected by the delay spread of the radio channel, whilst in CDMA systems the bursty property of the signals is rapidly degraded due to both the frequency dependence and delay variance of channels. This aspect of performance constitutes a fundamental difference between the three multiple access methods.

Virtually all practical TDMA and CDMA models have a FDMA component, even if this is not formally part of the design. A FDMA component is introduced as a result of the extent of the total bandwidths provided for mobile radio systems, which are some tens of MHz. The occupancy of such large bandwidths by individual TDMA or CDMA users signals would require large-bandwidth components to be used in the mobile terminals, which might be impractical to implement, and in the case of TDMA, very short burst lengths would be necessary to avoid excessive latency periods. Furthermore, in respect of resource management in multi-service and multi-operator environments, subdivision of the total available bandwidth into partial frequency bands introduces a necessary element of FDMA.

A key feature of cellular mobile radio systems is re-use of access channels. This means that the available frequency bands and/or time slots allotted to users in one cell in accordance with the multiple access scheme operating are also allocated to other users in other cells. The cells re-using the same frequency or time slot resources are separated from each other appropriately according to a re-use pattern. A group of cells not sharing the same resources is known as a cluster, and the number of cells in a cluster is the re-use factor, or cluster size. It is most obvious and practicable to re-use frequency, which means that different individual cells in a cluster utilise different parts of the available radio bandwidth. This can only be accomplished by using an FDMA element. Other user signals that are active at the same time and are using the same frequency band as a particular user give rise to co-channel multiple interference, and this is the limiting factor in a cellular mobile radio system. The sources of co- channel interference can be sub-divided into those located within the cell, which contribute to intra-cell co-channel interference, and those located in other cells, which give rise to inter-cell co-channel interference. The FDMA and TDMA schemes offer immunity from intra-cell co-channel interference, whereas in the case of CDMA, both intra-cell and inter-cell co-channel interference are present.

TDMA has a clear advantage. Transmission bursts in TDMA in a delay-varying radio channel are short enough to allow the signal to be considered of fixed delay, which means that relatively straightforward receiver and equaliser designs can be used. The sporadic nature of user signals in TDMA gives rise to near ideal signal separation at the receiver. This property of signal burstiness also reduces the need for power control. For a given data rate, transmission using combined F/TDMA access schemes have greater channel bandwidth requirements than for pure FDMA schemes. TDMA operation in conjunction with interleaving and forward error correction coding helps to combat the effect of fast fades in the radio channel. This feature is called time diversity. By allowing smaller or larger time slots to individual users, the data rate can be flexibly adjusted on demand. A difficulty with TDMA is that the timeslot model requires mutual synchronisation of the different users of a cell.

CDMA has two roots: a spread spectrum technique with single user detection, and a multi-user concept. In CDMA system designs having single user detection, all user signals from a cell except the desired user signal and all signals from other cells are seen as noise at the receiver. Therefore, in contrast to TDMA, the desired user signal is disturbed by both inter-cell and intra-cell interference. The standard approach to reception and detection in CDMA systems with single user detection uses the RAKE principle. In CDMA system designs with joint detection, all user signals in a cell are detected simultaneously by exploiting the CDMA codes used. Thus intra-cell interference is eliminated in CDMA systems using joint detection. An intermediate approach to detection between single user and joint detection is interference cancellation, which relies on first identifying the strongest user signal in the composite received signal, then reconstructing the contribution from the total received signal, and finally subtracting this contribution from the total signal. By this process, which is performed repeatedly, the effective interference to the residual user signal can be reduced. Compared to FDMA, CDMA also results, like TDMA, in an increased available user bandwidth, which equates to increased frequency diversity. This has advantages in the frequency selectivity of mobile radio channels. Another advantage of CDMA is that the multiple access interference experienced at each receiver input is generated by a larger number of signal sources than in non-CDMA systems. This larger interference diversity reduces fluctuations in interfering power at the receiver caused by both long and short term fading, and thus leads to lower probability of system outage. An advantage of CDMA having single user detection is flexibility, which arises from the fact that individual user signals are processed at the transmitters. In such systems, user signals can be switched on and off without coordination between transmitters, and data rates can be individually selected and changed. The price of this flexibility is increased intra-cell interference. In order to obtain sufficient capacity, the effect of intra-cell interference has to be off-set by accurate power control, voice activity monitoring and measures like soft handover (inter-cell handover) in conjunction with cell partitioning, which increases the amount of signalling data which must be passed over the fixed network. With CDMA using joint detection, the intra-cell interference is minimised. The accommodation of users with different and variable data rates, and the implementation of voice monitoring, requires a greater level of management in CDMA with joint detection compared with single user detection.

2.4 Standardisation

There are two different aspects of product standardisation:

- on the one hand, the definition of standards describing the technical characteristics that the product should conform to;

- and on the other, the certification or validation of products which establish that it adheres to the standard.

In world standards, the important bodies in the fields of telecommunications and information technology are:

- The International Standards Organisation (ISO), which operates in Geneva, and represents 89 countries through its national standards organisations. The standards published by ISO are for reference, and are not mandatory in the respective member countries;

- The International Telecommunications Union (ITU), also based in Geneva, has three principal bodies:
  1. The CCITT (Comité Consultatif International pour le Téléphone et le Telegraph) produces recommendations for information networks (X series), telephony (V series), facsimile (T series) and ISDN networks (S series) every 4 years;
  2. The CCIR (Comité Consultatif International pour les Radiocommunications);
  3. The International Frequency Register Board (IFRB), which allocates frequency bands in ITU member countries and adjudicates questions of radio interference caused by overlap.

The three principal bodies in Europe that enact these standards are:

- The Comité Européen de Normalisation Electrotechnique (CENELEC). Created in 1958, this committee is based in Brussels and represents 18 countries. These standards are mandatory for the member countries, who are required to establish corresponding equivalent national standards;

- The Comité Européen de Normalisation (CEN). This committee was created in 1961 and is also based in Brussels. The 18 countries represented are also members of CENELEC;

- The European Telecommunications Standards Institute (ETSI). This body represents 21 countries, was created in 1988 and is based at Valbonne in France. ETSI creates standards applying to European public telecommunications networks.

Certification bodies oversee the validation procedures and deliver conformance statements. There are 20 accredited certification bodies in Europe.

Standards guarantee users quality, technical performance and secure operation, and provide reference points based on several criteria. In addition, the standards open the markets to manufacturers who follow the standards, but also limit their liability in the case of damage. A European standard enlarges the territory that the product can be used in, both for the consumer and for the manufacturer, and each can thereby benefit. For the consumer, a standard simplifies life (for example, his or her credit card is accepted throughout the world because it is standardised), and for the manufacturer, the standard expands the market for the products, and therefore the opportunities.

2.5 Network Architecture

The ISO has established a conceptual architecture hierarchy for telecommunications networks which takes the form of seven logical layers. This hierarchy is called the Open System Interconnect (OSI). This is shown in Figure 2.3. Each layer is assigned a role, and carries out well defined functions. In a network node, a layer N communicates vertically with layers $N + 1$ and $N - 1$ through an interface, and by means of service requests, and exchanges information horizontally with a peer layer N located in another node. This is achieved using a protocol appropriate to layer N. Two adjacent layers communicate via messages called Protocol Data Units (PDUs). One of the main concepts of this architecture is that a layer of level N offers its services to the next higher layer; when a layer $N + 1$ requests a service from layer N, it can specify the value of certain parameters. There is thus a negotiation phase between the two layers.

The communication process between layers uses the principle of encapsulation. A message from layer N destined for the network is not modified by layer $N - 1$, but is inserted as data in the message to layer $N - 2$. Figure 2.4 illustrates this mechanism. On receiving a message from the network, the layers use an extraction mechanism to deliver the message destined for the layer above.

7	application
6	presentation
5	session
4	transport
3	network
2	link
1	physical

Figure 2.3: OSI 7-layer hierarchy

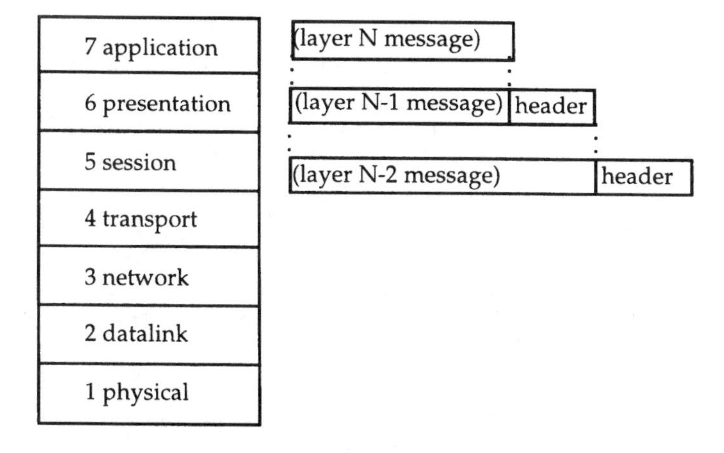

Figure 2.4: Message encapsulation

Consider a transaction in which the application layer of node A wishes to send a message to the application layer of node B. To do this, it transmits the message to its presentation layer, which inserts it in the data field of its own message, putting information for the presentation layer of node B describing the source, destination message type, etc. into the message header. Following this pattern, each layer adds its own descriptive message header down to the physical layer which sends out the message on the transmission medium towards node B.

Here, a message extraction process is invoked to restore the original message:

Physical layer
At its lowest level, the physical layer interfaces with the transmission medium, and at its upper level, with the datalink layer above. The physical layer deals with the characteristics of the physical means of transmission in the network. Software in the physical layer sends and receives information relating to the transmission bearer, the main function of which is to transport

binary data. It is at this level that the basic transmission characteristics are defined; transmission type (synchronous, or asynchronous), bit rate, transmission mode (analogue or digital), type of coding of digital data on the transmission bearer, and the type of protocol for point-to-point or point-to-multipoint dialogue with other network nodes. The physical layer exchanges data frames with the data-link layer.

Datalink layer

Situated between the physical layer and the network layer, the datalink layer receives data frames from the physical layer which it assembles into packets addressed for the network layer above. The principal function of the datalink layer is the detection of transmission errors in the frames arriving from the physical layer, so that these do not affect the operation of the higher layers. One of the most commonly used protocols is the ISO High Level Datalink Control protocol. One version of this is specified by CCITT under the name LAP B (Link Access Protocol –Balanced). These protocols use error detecting codes operating via the use of redundant bits added to the data frames to form Cyclic Redundancy Check (CRC) controls. Using a known algorithm, the receiving node is able to calculate the checksum of each received frame in order to compare the received and transmitted values. When these differ, errors have occurred, and the frame can be rejected. The datalink layer guarantees a residual undetected error rate conforming to the required threshold. The threshold can be set by the network user through the choice of detection algorithm. In local networks, the datalink layer arbitrates network access when there is competition between several users. The datalink layer also controls the addressing of transmitting nodes, and is able to associate logical as well as physical addresses to them. The traffic flow between two nodes is also sometimes managed within the datalink layer.

Network layer

The network layer supervises the network. It continuously monitors the terminals connected to the network and registers their addresses in order to keep an up-to-date table of terminals and their status (in service, or out of service), which allows a routing table to be maintained. To carry out this function, the network layer sends out special frames, called signalling frames, as distinct from user data frames. Using the routing table, the network layer is able to determine the routing of packets through the network, i.e. it defines the intermediate nodes through which the packets must travel in order to reach their destination when the latter is not known to the transmitting node. Routing can take place in connection-oriented networks, where all the packets follow the same physical path but in a connectionless, or datagram, service, a routing is allocated individually for each of the transmitted packets. When offered by the network, the network layer re-orders the arriving packets so as to maintain the correct sequence. Flow control is another function of the network layer.

Transport layer

The transport layer marks the boundary between the physical elements in a network and the logical ones, and provides a communication service to the higher layers. The transport layer offers a transparent transfer of information between two session entities, in five service classes:

- Class 0: This is the basic transport layer service. Errors created in the network are detected but not corrected. This class meets the needs of the Teletex application, for example;

- Class 1: This class provides a reduced flow rate to cater for re-transmission upon error detection;

- Class 2: This class provides multiplexing of several transport connections in one network connection. Flow control is available. There is no re-transmission to recover from errors;

- Class3: This class offers multiplexing, flow control and re-transmission on errors flagged by the network layer. It represents the sum of the services offered by classes 1 and 2;

- Class 4: This is the most complete class. It provides flow control, retransmission on error detection and multiplexing of several transport connections on one network connection.

The most common transport layer protocols are: ISO 8072, CCITT X224 and TCP (Transmission Control Protocol). The class of service is dictated by the requirements of the application and the quality of the network.

Session layer

The session layer synchronises the dialogue between two applications and controls the services offered by the transport layer. It identifies the applications but it does not manage the transmission of data which is assumed to be perfect. The session layer permits data to be typed and structured. Various standards exist for this layer, ISO 8326, 8327 and CCITT X.215, X.225. Not all of the functions of the standards are used, but there are communication profiles defined by statutory bodies, user forums and equipment manufacturers. Options are negotiable at the setting up of a connection by the presentation layer such as:

- the maximum size of data blocks;

- the type of user dialogue (half duplex, full duplex, expedited data).

Presentation layer

The presentation layer manages the description of data and the syntax of data structures. For point-to-point logical connections, the presentation layer presents the data in the same way at the two extremes of the network, and this allows environments to be defined. The presentation layer 'negotiates' the presentation of data to applications, using an exchange syntax based on a language known as Abstract Syntax Notation One (ASN.1). The ASN.1 language is defined in CCITT recommendation X.409 (ISO standard 8824). This language allows a distinction to be made between the information exchanged by peer application entities and their representation by means of a set of coding rules known as Basic Encoding Rules (BER) (ISO standard 8825). Just as for ASN.1, these rules are also standardised. These rules are very useful for describing complex information. On the one hand, they allow the description of information with the aid of the formalism of high-level languages, which facilitates manipulation by the designer; and on the other, they offer the programmer a means of rapid machine code translation.

Application layer

The application layer is not a true functional layer of the model, since it does not deal with operational processes. It provides a front-end framework aligning tasks that need to be interfaced and creates a user interface to the communications process. The application layer defines the interactions between the user and the communications system.

The most commonly used types of application offering service to terminals are the File Transfer Access Method (FTAM) used for manipulating files, the File Transfer Protocol (FTP) for exchanging files, and the Product Definition Interchange Format (PDIF) for exchanging messages supporting the transfer of documents, text or graphics across the network.

The GSM Standard

3

3.1 Scope of the GSM Standard

The GSM network provides subscribers with three types of service. These are:

- bearer services;
- teleservices;
- supplementary services.

Figure 3.1 shows the main elements in a radio telephone system. Subscribers gain access to the network through their handset terminal, also known in GSM terminology as the mobile station. The mobile station is both a wireless telephone and a data terminal which receives messages from the network, and the Base Transceiver Station (BTS) is the network terminal equipment interfacing to the mobile. The BTS consists of a group of fixed transmitters and receivers through which it transfers messages to the mobile terminals present in the cell that it controls. It uses different radio channels according to the type of information to be transferred – user data or signalling – and according to the direction of information flow – subscriber-to-network or network-to-subscriber. In the network, after the Base Transceiver Station is the Base Station Controller (BSC); it communicates with one or more BTS. These units concentrate traffic from the base stations and act as gateways to the network sub-system. The base station controllers are served by Mobile

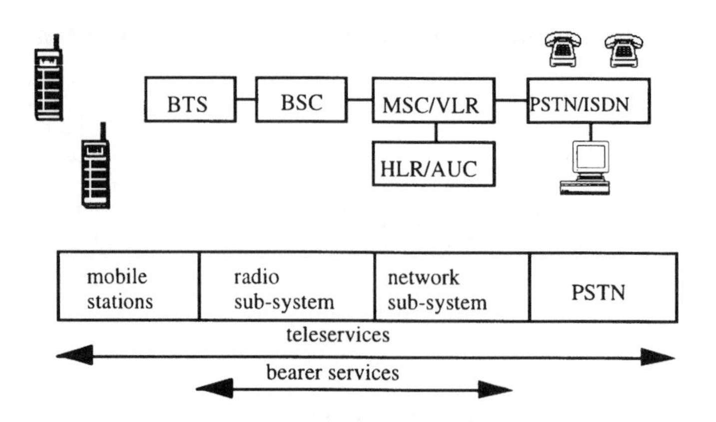

Figure 3.1: GSM network structure

Switching Centres (MSC) which connect the GSM network to the PSTN and ISDN public switched networks, and provide the interface between the GSM network data base and the radio sub-system. These databases allow the access rights of subscribers to be checked, and also keep track of the location of subscribers. The data bases are the Visitor Location Register (VLR), the Home Location Register (HLR) of the Switching centre on which the subscriber is based, and the Authentication Centre (AUC). The VLR database stores information relevant to users who are in transit.

The HLR of a GSM network subscriber is a data bank. It contains the sources of information relating to this subscriber, particularly the profile of his subscription. When this subscriber enters the network, or when he requests access to a service, network equipment which can monitor the validity of the requested privileges interrogates the subscriber's HLR. A subscriber's KLR contains permanent information. On the other hand, a VLR holds temporary dynamic information relating to a mobile station.

3.2 Bearer Services

The bearer services provided by GSM allow end to end transfer of information within the network and are defined by the technical attributes of the network perceived by the user at the point of access. The GSM standard identifies three sorts of attribute:

- information transfer;
- access;
- general attributes.

3.2.1 Information Transfer Attributes

Information transfer attributes describe the data transfer options in point-to-point and point-to-multi-point networks.

There are two types of attribute: primary and secondary. Primary attributes define a service category, and secondary attributes define the particular service within a category. Table 3.1 provides a list of these attributes. Attributes 1–4 are primary attributes, and the rest are secondary attributes. Attribute 1 – transfer mode – characterises the type of data transmission network that a subscriber can connect to. The network is either circuit switched, like telephone networks, or packet-switched. The type of data to be transmitted can impose a limitation on the application or on the transmission quality of the data. The call set-up mode also characterises the network with which data transfer is sought; either with or without a permanent circuit[1]. In a connection-oriented call, the sender exchanges signalling messages with the network before transmission takes place to find out if the destination node is free. Local networks are generally of the connectionless type. In addition, the communication configuration characterises certain networks in which information can be broadcast from one terminal to a group of terminals. Attribute 7 defines the data transmission protocol between nodes, and the status of each of the parties (whether data sources or sinks).

Table 3.1 List of information transfer attributes

1	Data transfer mode (circuit, packet)
2	Data rate
3	Information type (data, speech)
4	Information structure
5	Call set-up mode
6	Data transfer mode (point to point, multipoint, broadcast)
7	Unidirectional/bi-directional (symmetric, asymmetric)

3.2.2 Access Attributes

Table 3.2 lists the access attributes, which describe the means of accessing functions and supplementary services of a network.

[1] In a network of connection-oriented circuits, a call has three phases: call set-up, data transfer and release. In a connectionless circuit, there are no set-up and release phases. The originating node transmits data without first establishing the presence of the destination node.

Table 3.2 List of access attributes

1	Access channel and bit rate
2	Access protocol (data, signalling)

3.2.3 General Attributes

Table 3.3 lists the general attributes, which refer to the whole set of supplementary services.

Table 3.3 List of general attributes

1	Guaranteed supplementary services
2	Quality of service
3	Interworking options
4	Operational and commercial

A service is identified by the access and general attributes, which specify in detail the bearer service. Figure 3.2 is a schematic of the data transmission model.

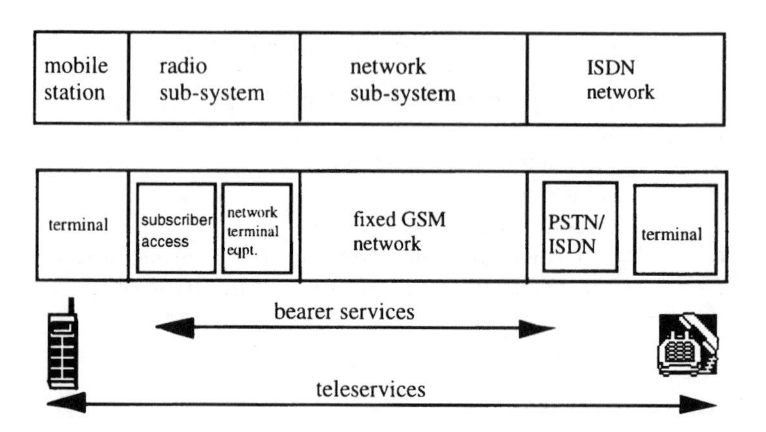

Figure 3.2: Data transmission model

GSM provides bearer services without restrictions on the type of user data to be transported, and the services convey that data without modification across the entire GSM network. This mode of transport ensures time-sequence integrity by setting up a single virtual circuit path. User data and network signalling are transported in different communications channels in the GSM network. The bearer services available in GSM are suitable for a wide variety of applications, such as speech transmission, X.25 packet network access, multimedia data transfer, paging, etc.

Table 3.4 lists the GSM bearer services available to end-user applications. Both connection-oriented and connectionless data network transport modes are possible with GSM. With connection-oriented transport, the radio protocol provides extra protection against transmission errors by message acknowledgement.

Table 3.4 GSM bearer services

Asynchronous duplex data transmission	300–9600 bit/s
Synchronous duplex data transmission	1200–9600 bit/s
Synchronous access to a PAD	300–9600 bit/s
Synchronous duplex packet transmission	2400–9600 bit/s
Alternate speech and data	2400–9600 bit/s

Table 3.5 lists the modems currently supported by GSM.

Table 3.5 List of GSM modems

Type	Data rate, bit/s
V21	300
V22	1200, 2400
V23	1200/75
V26	2400
V32	4800, 9600

3.3 Teleservices

Teleservices are the functional telecommunications operations made available by the network to subscribers. They make use of the options offered by the bearer services and allow user-to-user communication within the context of an application.

Telephony is the most important of the teleservices and allows different types of communications. These are:

- communication between two mobile terminals;
- communication between a mobile and a fixed terminal, across any number of networks.

A call to the emergency services is generated automatically when the user presses a dedicated key on his handset. In addition, group 3 fax is offered, by using an adaptor.

GSM also offers short (140 characters maximum) alphanumeric message transactions, in both send and receive modes. This service can be invoked on all terminals capable of sending alphanumeric messages to a GSM terminal. A reception acknowledgement confirms delivery of the message to the destination. This marks an improvement compared to normal radiopaging. This service is possible in both point-to-point and point-to-multipoint modes. Table 3.6 lists GSM teleservices, and Table 3.7 lists attributes associated with teleservices that are different from bearer service attributes.

Table 3.6 GSM teleservices

Information type	Service offered
Speech	Telephony
	Emergency calls
Data	Point-to-point paging
	300 bit access
	1200 bit access
Text messages	Transmission of short alphanumeric messages (max. 140 bytes)
Graphics	Group 3 fax

Table 3.7 Teleservice attributes

Layer 1 and 2 attributes	
1	Information transfer mode
2	Information data rate
Layer 3–7 attributes	
3	Type of data coding (telephone, audio, text, fax, videotext, etc.)
4	Transport layer protocol functions
5	Session layer protocol functions
6	Presentation layer protocol functions
7	Application layer protocol functions

3.4 Supplementary Services

Supplementary services support and augment the other services. They are numerous, and the following list is not exhaustive:

- caller identifications;
- call forwarding – unconditional, busy, no reply, network congestion;
- call waiting;
- call hold;
- charging advice;
- call barring – outgoing, incoming, outgoing international;
- voice messaging;
- dual dialling;
- conference calls;
- transfer of call in progress;
- short code dialling;
- closed user groups;
- call-back when free.

3.5 Multiplexing Techniques

Multiplexing is a technique for sharing a scarce transport medium resource by use of a fixed partitioning between several users in a constant sequence. In order to optimise frequency use and to provide the maximum number of transmission channels, the GSM standard has been developed with two simultaneous multiplexing techniques: Time Division Multiple Access (TDMA) and Frequency Division Multiple Access (FDMA).

3.5.1 TDMA

In temporal multiplexing, the resource to be shared is time. The frame is a constant time-interval sequence, and this forms the basic transmission unit. Multiples of the basic frame can be assembled into multiframes, superframes and hyperframes. A frame is composed of a number of time intervals, known

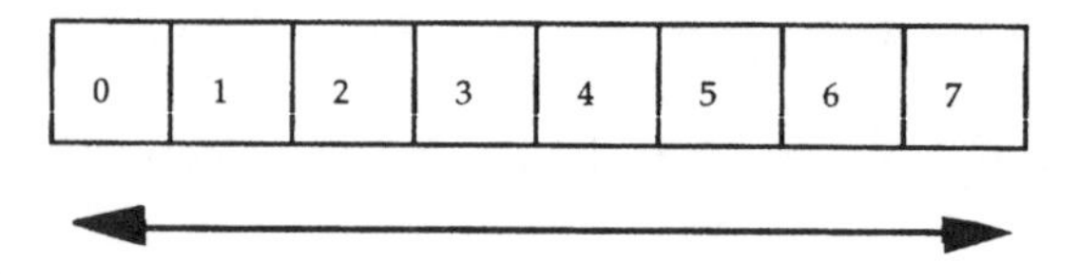

Figure 3.3: Structure of eight time slot TDMA Frame

Time slot 7	Communication channel No. 7
Time slot 6	Communication channel No. 6
Time slot 5	Communication channel No. 5
Time slot 4	Communication channel No. 4
Time slot 3	Communication channel No. 3
Time slot 2	Communication channel No. 2
Time slot 1	Communication channel No. 1
Time slot 0	Communication channel No. 0

Figure 3.4: Relationship between time slots and communication channel

as time slots, and the GSM standard divides the frame into eight time slots, as shown in Figure 3.3. A correspondence is established between a time slot and a communication channel, as illustrated in Figure 3.4. The multiplexing advantage comes from multiplying by eight the number of transmission channels available per unit time. Each of these communication channels transports point-to-point data in either one direction or another, mobile-to-network, or network-to-mobile.

3.5.2 Frequency Division Multiplexing

Frequency division multiplexing shares available frequency bands into 124 sub-bands for each direction of transmission. Each of these sub-bands constitutes a separate physical radio communication channel, with FDMA providing 124 simultaneously usable radio channels. The channel is identified by the central frequency of the band, known as the carrier frequency.

Figure 3.5 shows the relationship between the two multiplexing techniques in GSM. Each time slot in a frame corresponds to a carrier frequency.

In practice, the correspondence between a time slot and a carrier frequency is not permanent, and this complicates matters. However, the simple relation-

Time slot 7	F8
Time slot 6	F7
Time slot 5	F6
Time slot 4	F5
Time slot 3	F4
Time slot 2	F3
Time slot 1	F2
Time slot 0	F1

Figure 3.5: The relationship between TDMA and FDMA multiplexing

ship described above aids understanding. The relationship linking a communication channel and a time slot is fixed but the relationship linking a communication channel and a frequency is dynamic. At the beginning of a call a pseudo-random selection algorithm is invoked to allocate frequencies to logical communication channels in each time slot. In this way, a communication channel corresponding to a given time slot uses a frequency determined by this selection algorithm.

There are several reasons for such sophistication: firstly, in order to guarantee confidentiality; and, secondly, to protect a physical communication channel against the effects of radio interference which may exist in any given

Table 3.8 Speech processing in GSM

Disturbance	Physical	Logical
		Digitisation
		Coding
		Encryption
	Modulation	
	Emission	
Scrambling		
Interference	Transmission	
Fading		
	Reception	
	Demodulation	
		Decoding
		De-cryption
	Synthesis	
	Speech	

frequency band. Indeed, with this method, only a single time slot is perturbed if there is interference on a particular carrier frequency at a given time and in this event the next data packet, which is being transported on a different carrier frequency, will not be affected by the interference. In addition, if a receiver has difficulty in detecting a particular frequency, this will only affect the communication channel for one data packet.

Table 3.8 shows the main processing applied to speech transmitted between source and destination and the type of disturbance that it might be subjected to.

3.5.3 Code Division Multiplexing

Code division multiplexing is best known by its abbreviation CDMA (Figure 3.6). This digital transmission technique has a military origin. With this method, the axis of channel separation is the method of coding the information. All channels use the whole bandwidth at the same time. The coding algorithms generate codes having a low probability of being detected and a low probability of interception. Two different channels use orthogonal codes. The radio signal produced by a transmitter simulates noise. One illustration of the use of this type of coding is that of a reception at an embassy where all the guests speak simultaneously and where only two individuals speaking the same language understand each other.

Figure 3.6: An example of CDMA multiplexing

The advantages with respect to the other techniques are:

- There is no plan for frequency re-use;
- The number of channels is greater;

- Protection against fading of the signal;
- Better protection from interference;
- Optimum utilisation of the bandwidth;
- Confidentiality of communication is well protected, the two correspondents are the only ones to know the coding algorithm.

The disadvantages are as follows:

- Transmission power must be finely controlled;
- The other channels are noise sources.

The satellite Global Positioning System (GPS) and GLObal Navigation Satellite System (GLONASS) use this method of multiplexing. CDMA is a serious candidate for third generation networks under development.

3.6 GSM Requirements

GSM standards are drafted by ETSI. In order to define the different needs of the main participants in GSM communications, ETSI has published over 140 technical specifications covering GSM and DCS 1800 systems. [DCS 1800 is an ETSI standard for personal communications networks (PCNs).] There are three principal categories of standard. These are:

- GSM at 900 MHz;
- DCS at 1800 MHz;
- both groups (GSM 900 and DCS 1800).

The framework for ETSI standards is built around the following:

- the user;
- the network operator;
- the manufacturer;
- the regulating body governing network operation.

ETSI has established the requirements of each of the above groups that have a stake in GSM systems, and has established the standards to meet their needs.

3.6.1 User Requirements

Users look upon the radio telephone as a communications tool that must be simple and comfortable to use, and marketed at a reasonable tariff. Users therefore require user-friendly operation of the subscriber equipment and of the services offered by the network, and ETSI has focused on the following user requirements:

- a speech quality comparable to that of a conventional telephone;
- call privacy;
- a wide network coverage;
- messaging services;
- data services;
- lightweight, compact and ergonomic handsets;
- reasonable cost of access;
- reasonable usage tariffs;
- high service availability;
- international roaming (the ability to use the service within foreign networks, other than the one within which the subscription is based).

3.6.2 Network Operator Requirements

The network operator considers the network as an investment that must be affordable, must have a degree of future-proofing and must be upgradable. A network operator is, above all, a provider of services to the subscriber. His network is both a working tool and the mainstay of his commercial activity. The network qualities that he needs are:

- optimum resource utilisation (radio frequencies and transmission capacity);
- a high availability;
- simple and efficient operation;
- simple, effective and reliable means of identifying subscribers and terminals;

- a large number of subscribers;
- standardised equipment;
- flexible standards;
- several equipment manufacturers;
- a reasonable infrastructure cost.

3.6.3 Manufacturer Requirements

The manufacturer makes the equipment. He needs standards that provide the following:

- a stable definition of the functionality of the products;
- a clear definition of the constraints;
- a single product certification authority;
- as wide a market as possible.

3.6.4 Requirements of the Regulating Body

The aims of the regulation authorities are:

- to open up radio telephone access to the whole of the population;
- to open up this type of service to free competition between operators;
- to standardise equipment in order to open up the supply market to all manufacturers;
- to standardise subscriber equipment;
- to use the full potential of the limited number of frequencies allotted to GSM;
- to see radio telephony deployed throughout the country of their administration.

The regulation authorities wish to act equitably on behalf of all the 'players' in GSM networks by adopting liberal policies aimed at guaranteeing options in the service offering, and a free choice for consumers.

GSM Network Infrastructure

4

4.1 Overview of Network Infrastructure

Figure 4.1 shows a schematic view of the equipment functions making up the GSM network. The network is characterised by its equipment elements and the interfaces that link them together. All these elements provide standardised services.

4.1.1 Network Equipment Functions

The equipment types in the network are:

- the subscriber terminal, known as the Mobile Station (MS);
- the base station, which is a radio receiver and transmitter, known as the Base Transceiver Station (BTS) – it links the mobile stations to the fixed network infrastructure;
- the Base Station Controller (BSC), which controls a group of base stations;
- the grouping of base station and base station controller which is known as a Base Station Subsystem (BSS);

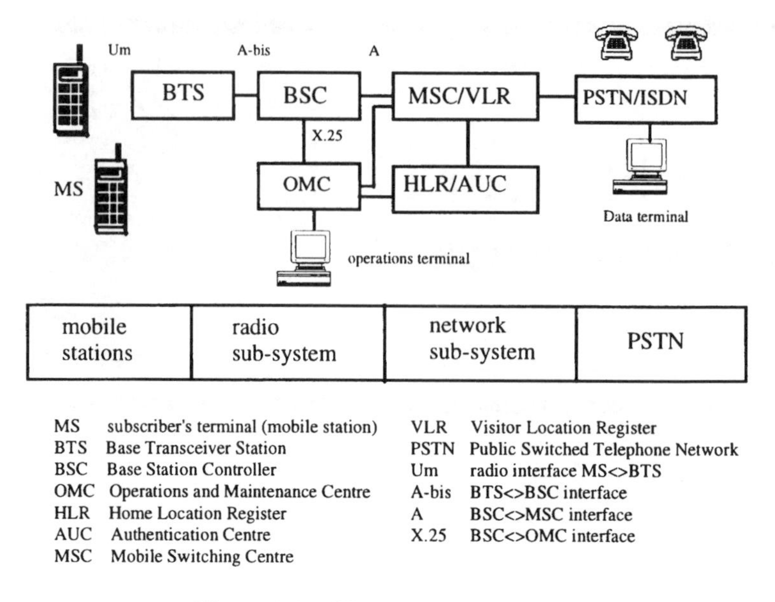

Figure 4.1: GSM network description

- the network switch, or Mobile Switching Centre (MSC), which provides connections to the external PSTN or ISDN networks;

- the Visitor Location Register (VLR), which is a database recording temporarily the details of subscribers who are transiting through the network;

- the register associated with the home location of subscribers, known as the Home Location Register (HLR), which is a database containing the basic reference identity and user details of the subscriber;

- the subscriber AUthentication Centre (AUC), which is a secure database where the confidential subscriber access codes are stored and controlled;

- the grouping of the switch, the HLR, the VLR and the AUC which constitutes the Network Sub-system (NSS);

- the Operations and Maintenance Centre (OMC), which is responsible for the logistical and technical operations of the network.

4.1.2 Network Interfaces

The interfaces are also important parts of the network, supporting dialogue between the different equipment elements, and facilitating their interworking. The standardisation of the interfaces guarantees interoperability of different makes of equipment. As a result, ETSI has standardised all the following interfaces:

- the 'Um' radio interface, which is situated between the mobile station and the base station MS ⇔ BTS, and is the most important interface in the network;
- the A-bis interface, which links a base station to its controller BTS ⇔ BSC via a digital circuit; the support is a wired PCM link;
- the A interface; which exists between the controller and the switch BSC ⇔ MSC and is linked via a 64 kbit/s PCM link;
- an X.25 link, which connects the controller to the operations centre BSC ⇔ OMC. The link support is provided by a data transmission network;
- the interface between the switch and the public network, which is defined by the CCITT signalling system No.7.

4.2 GSM Network Equipment

4.2.1 The Base Station (BTS)

The cell is the basic element of service coverage in a zone, and the BTS ensures the radio coverage of the network cell area. It provides an entry point to the network for subscribers who are present in the cell, allowing them to make and receive calls. The base station can handle up to eight simultaneous communications, the maximum number being fixed by the time division multiplexing used. The area of the cell varies significantly between urban and rural districts. In urban areas where the traffic density is high, the cell size is small in order to increase the capacity per unit area. In these situations the radius of a cell can approach the lower limits (200 m) imposed by the cost of infrastructure and the propagation conditions. On the other hand, in rural areas, the traffic density is much lower, and the size of cells is therefore much larger (~30 km), and it is the transmitter power that determines the limits. A base station is essentially a transmitter – receiver set, and is itself an element in the communication chain. Base stations are operated either locally, if necessary, or remotely via the BSC.

4.2.2 The Base Station Controller (BSC)

A controller governs one or more base stations and carries out a number of operational and telecommunications functions. For traffic coming from the base stations, it acts as a concentrator, and for traffic arriving from the switch, it acts as a router towards the destination base station. In its network operation functions, the controller firstly relays alarm and performance statistics sent out by the base stations and destined for the operation and maintenance centre and secondly acts as a database for configuration software downloaded to the base stations from the operations centre via the controller. It holds and sends this data as requested by either the network operations centre or by the base station as it starts up. In its network operations role, the controller manages the radio resources for a zone comprising the group of cells for which it is responsible. As a consequence, it assigns the frequencies that can be used by each of the base stations. The controller also oversees the handovers when a mobile station crosses the boundary between two cells. To achieve this, it informs the cell that is due to take over the subscriber's call and sends him the necessary information. In addition, the controller informs the HLR database of the new location of the subscriber. The controller is one stage in the communications chain linking the subscriber's terminal during the teleservice communication phase, or during the search for the subscriber when a call arrives for him from the switch. The BSC is the only part of the radio sub-system that can be directly operated remotely from the operations and maintenance centre. Technical control of the base stations is carried out via the controllers.

Figure 4.2 shows schematically the BSC and the network elements it is attached to, a group of three base stations, an operations and maintenance centre and a switch.

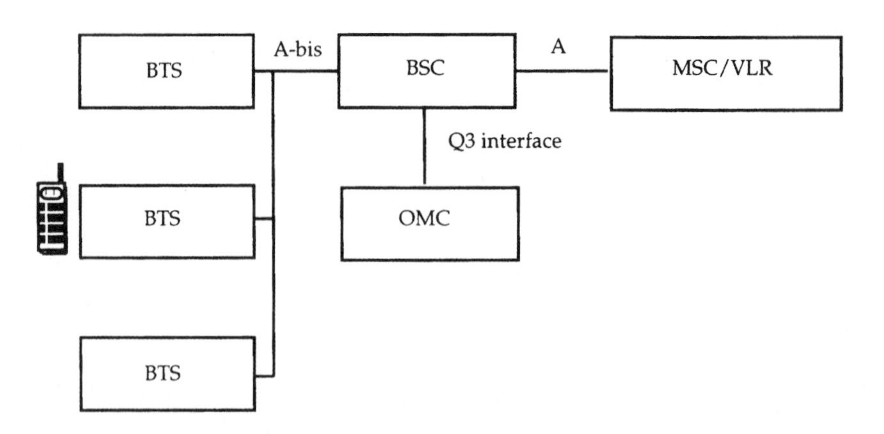

Figure 4.2: Radio sub-system interfaces

4.2.3 The Switch (MSC)

The switch connects the radio telephone network to the public telephone network, and fulfils the specific requirements arising from mobility, inter-cell transfer, and the management of visiting subscribers who belong to outside networks. The switch is based on the automatic ISDN switch design, adapted to include the extra necessary functionality for cellular networks. In addition, it is an important network node, giving access to the network databases and to the AUC which checks subscriber access rights. It aids the control of mobility of subscribers, and thus of monitoring their location in the network, but is also involved in the provision of teleservices offered by the network, as well as telephony, supplementary services and message services.

4.2.4 The Home Location Register (HLR)

The HLR is a database for subscriber data. A cellular network may possess several of these registers, depending on the equipment capacity, the reliability, and the operations policy of the network operator. The data recorded in them includes the subscription details, including the options taken up and the supplementary services available to the subscriber. Together with this fixed data, there is variable information describing the last known location of the subscriber, and the status of his terminal (in-service, out of service, making a call, ready to receive a call, etc.). The HLR distinguishes between data relating to the subscriber and that relating to the terminal. A subscriber may use the terminal of another customer without any problems of billing, because subscribers are identified by information contained on an intelligent personal subscriber card, possessed by all subscribers and known as a Subscriber Identity Module (SIM), which contains a microprocessor. When a subscriber makes use of the service, some of the information contained on the SIM is transmitted to the HLR database, which recognises the subscriber. In this way, the network distinguishes between the subscriber and the terminal. Temporary information about the location of the subscriber is continuously updated, and the messages to be delivered to the subscriber and the telephone number of a terminal diverting calls are stored in the HLR. The dynamic information is particularly useful when the network is setting up a call to the subscriber. At the beginning of call set-up, the network starts by interrogating the HLR to establish the last known location of the subscriber, his terminal status at that time, and the date of this information, before taking any action.

The major difference between the fixed network and a radio telephone network is the mobility of the subscriber's telephone. In a fixed network, the telephone number is associated with the address of a fixed terminal at a given site. The system can therefore quickly determine the path between the source of a call and its destination. This method is inadequate in a mobile network, where the route is arrived at by successively interrogating databases to locate the destination, before setting up the call. The HLR also contains the secret encryption key associated with the subscriber, allowing the network to identify him. This key is stored in a coded format that only the network AUC is able to decipher.

4.2.5 Authentication Centre (AUC)

The AUC is a database that stores confidential information. As a consequence, it is located in accommodation that only authorised personnel can enter. Before anyone may access the database, they must enter a password. Furthermore, the data stored in the database is recorded on to the physical memory in a coded form. The AUC controls the rights of usage of the network services held by all subscribers. This verification is carried out on each occasion on which the subscriber uses the network. Such checking is aimed at protecting the network operator as well as the subscriber, and it is indeed important for the network operator to know unambiguously who has used the network in order to be able to bill him for the service used. Moreover, the accurate identification of users protects all subscribers against the fraudulent use of their accounts, and thus avoids their being billed for calls that they did not make. Since the network prevents fraudulent use, there are no grounds for contesting the bill. Subscribers know that they are paying for services actually used.

The identification proceeds in two stages. The first is local. When the subscriber switches his terminal on, he has to enter his identity in the form of an electronic signature by typing in a confidential code on the keyboard. This is verified by the microprocessor in his SIM card, which has been previously inserted into his terminal. Once the code has been entered and acknowledged, he can use the terminal. The second stage of identification occurs when the subscriber uses a network service. In this case the network firstly asks the terminal to supply the subscriber identity, which is his subscriber number. Secondly, the network asks the subscriber to prove his identity, using.an algorithm stored in read-only memory on the card. A copy of this algorithm is also stored at the AUC. In this way, the secret algorithm is never transported over the network. Only the result of the calculation performed with this algorithm circulates there in a coded form. The AUC authenticates the subscriber by comparing the submitted code with the correct one.

Although different from a normal telephone number, the subscriber number is not confidential, and someone attempting to defraud may know the number of one of his relatives. However, if he masquerades as someone else, he will be unable to prove his identity when asked by the AUC. The AUC then indicates this outcome and blocks his use of the network. When a subscriber is authenticated, the network interrogates the HLR to check the network options subscribed to, and the subscriber's access rights. If he has valid rights, the subscriber is given the access he is seeking.

4.2.6 The Visitor Location Register (VLR)

The VLR is a database associated with the switch (MSC). Its purpose is to store temporary information about subscribers moving through the network. This monitoring is important because, at any time, the network must know the whereabouts of every subscriber present; that is, in which cell they are located. In the VLR, the subscriber is described by a particular location identifier. The network needs to know this information, which is fundamental to being able to route a call or to set up a call requested by one visiting subscriber to another. Since the key feature of GSM for users is mobility, it is necessary to continuously locate the subscribers present in the network and to follow their movements. On each occasion on which a subscriber changes from one cell to another, the network needs to update the VLR of the network visited, and the HLR of the subscriber. This leads to continuous dialogue between the network databases. The updating of the HLR is important in order to process inward-bound calls. When the network attempts to contact a subscriber, it starts by interrogating the HLR to ascertain his last recorded location. It then interrogates the VLR where the subscriber should he registered to check that he is there. Following this, the network is in a position to map the route between the caller and the called party, and to connect the call.

The NSS is made up of the following elements:

The HLR, the AUC, the MSC, and the VLR. Its roles are call control, mobility management, and the management of supplementary services and messaging.

4.2.7 Call Routing

Two types of call are described in order to illustrate the interactions between the network elements. These are:

- a call from a GSM subscriber to a PSTN/ISDN subscriber;
- a call from a PSTN/ISDN subscriber to a GSM subscriber.

To make a call, the GSM network user dials a number. His call goes to the BTS of his cell, and then traverses the BSC before arriving at the network switch. At this point, the authorisation and usage rights are checked. Following this, the switch (MSC) transfers the call to the public network and requests the base station controller to reserve a channel for the impending call. When the called party answers the phone, the call is established.

When a PSTN/ISDN subscriber calls a GSM network subscriber, the processing is different and more complex. There is no initial control exercised by the network when a PSTN/ISDN subscriber dials a number, except for a potential outgoing call-barring function. The called number is first analysed by the subscriber's local network switch, and then a call request is directed to the GSM network to interrogate the HLR of the called number in order to locate the called mobile terminal. The HLR of the mobile network subscriber is the database holding the information about him and about his terminal status (free, busy or out of service). When the called party is free, the network interrogates the VLR in which he is registered in order to find out the cell and the base station controller of the zone that is able to make the connection to the called number. In order to ring the bell of the called terminal, the BSC of the relevant zone broadcasts a 'ring telephone' message through all the BTS stations in the zone. The called terminal listens to the network and, when it recognises its own number, it activates the bell/buzzer of the terminal. It is only after the phone is answered that the network finally allocates the reserved resources to the call. At the same time, the HLR and VLR update the status of the subscriber.

4.2.8 The Operations and Maintenance Centre (OMC)

The OMC is the control centre for operation and configuration of the network. It includes subscriber administration and technical management of the equipment. Administrative and commercial network management involves the provision, configuration and cessation of service, and billing. Much of the configuration management requires interaction with the HLR database. Commercial management processes gather statistics in order to build up a pattern of subscriber preferences and expectations. According to the information gained, the tariff may be varied in order to spread the traffic out during the day, when the traffic density for the popular services is highest. Technical network management seeks to ensure the availability and optimum configuration of the network hardware. The main areas of OMC activity are the supervision of equipment alarms, the rectification of misoperations, the control of software versions, performance management and security management. Most of the management tasks are instigated remotely from the network equipment by network messages, using a data network separate from the GSM network.

4.3 Radio Transmission

4.3.1 Physical Channels

The GSM Um interface utilises two multiplexing techniques; frequency division multiplexing (FDMA), and time division multiplexing (TDMA). Frequency division multiplexing divides each frequency range into 124 channels of 200 kHz width. The ranges used are; 890–915 MHz for terminal-to-base station transmission, and 935–960 MHz for base station-to-terminal transmission. This creates 124 parallel duplex communications channels, with each direction of transmission having its own reserved physical channel. Time division multiplexing shares the use of each frequency transmission channel into eight different communications circuits. A radio channel operates at a certain bit rate D per unit time. This bit rate is divided by eight to transmit successively the eight communications channels, each one having a bit rate of $d = D/8$, and each communications channel occupies a time slot of duration 577 μs, the sum of eight time slots constituting a frame, which is the basic time unit. A frame lasts for 4.615 ms in GSM.

Time multiplexing optimises the transmission capacity of a channel. In telephony, the average bit rate is small because of the numerous silences in speech, and because only one person is normally speaking at any one time. During a conversation, two successive portions of speech are transported as two successive frames, and these are separated by a duration of 4.615 ms. However, the sound reconstruction process preserves the continuity of speech.

The GSM standard defines a hierarchy of multiples of a frame: the multiframe, the superframe and the hyperframe. Table 4.1 lists this frame hierarchy and Figure 4.3 shows the arrangement schematically.

Table 4.1: GSM frame hierarchy

Multiframe_{51}	= 51 TDMA frames	
Multiframe_{26}	= 26 TDMA frames	
Superframe_{51}	= 51 multiframes_{26}	= 1326 frames
Superframe_{26}	= 26 multiframes_{51}	= 1326 frames
Hyperframe	= 2048 superframes	= 2715648 frames

A frame is divided into eight time intervals of 577 μs duration, and each interval carries a communications channel in which a message element, called a packet, is periodically transmitted. The packet is a structured collection of binary information elements (bits).

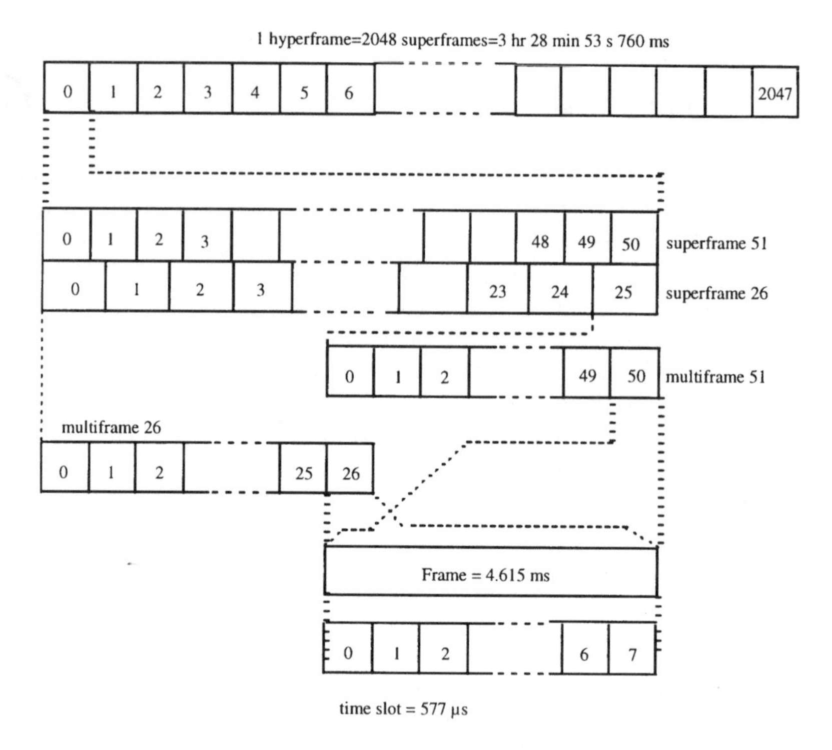

Figure 4.3: Description of the TDMA frame hierarchy

4.3.2 Packet Burst Structures

The standard defines four types of packet bursts:

- access bursts;
- synchronisation bursts;
- normal bursts;
- and frequency correction bursts.

For each of these types, a particular structure has been designed and a precise purpose is defined, as their names suggest. Figure 4.4 shows the structure of the burst types.

It can be seen from Figure 4.5 that each of these burst types has a four-zone structure.

The burst structure is a container that carries useful information, and is preceded by a Tail Bit (TB) zone. The end of the burst has a Guard Period

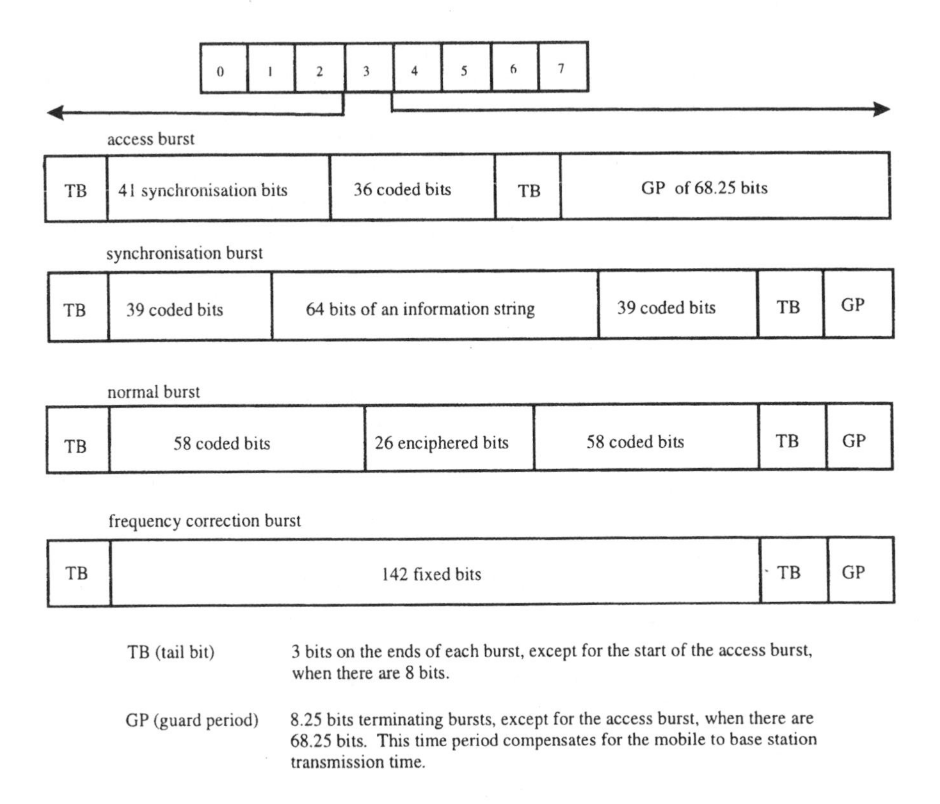

Figure 4.4: GSM packet burst structure types

A	B	C	D
TB	data field	TB	GP

Figure 4.5: Basic GSM packet burst structure

(GP) which serves to compensate for the duration of transmission of the bursts. From one packet burst to the next, this time duration is variable for a receiver, because either the transmitter or the receiver could have moved in the interim. The time slot interval of 577 μs is equivalent to the duration of 156.25 bits. In other words, the overall transmission bit rate is 270 kbit/s. However, the maximum usable bit rate for the subscriber is only 13 kbit/s.

Access packet burst

All mobile terminals make contact with the network through the transmission of an access packet burst in the dedicated access channel. This is the smallest of the packet bursts, transporting 77 bits (41 synchronisation bits and 36 data bits), and it has the longest guard period of 68.25 bits, equivalent to 252 μs. This guard time allows communications to be set up with mobile terminals which are up to 35 km from the BTS. The network continuously calculates the time of flight of a packet burst and uses this to determine the precise moment at which the mobile terminal should start transmitting in order to compensate the delay due to the radio propagation between the mobile and the base station. The transmitters in the network all have their clocks synchronised by a timing marker broadcast by the base station.

Synchronisation packet burst

The synchronisation packet burst transports 142 bits, containing 78 data bits for the mobile terminals, conveying information about their location in the network (base station identity, cell identity and zone identity), and their network access frequency.

Normal packet burst

The normal packet burst transports 142 bits of data. The central 26 bits are a training sequence, used for setting receiver parameters: two bits indicate the type of channel usage (data or signalling), six defined bits are reserved for the ramping of the signal amplitude to its quiescent level, and a TB zone, of duration 8.25 bits, is also included. This packet burst type transports call data.

Frequency correction packet burst

The frequency correction packet burst transports 142 data bits from the base station to signal frequency adjustments.

GMSK modulation

GSM uses Gaussian Minimum Shift Keying (GMSK) modulation. This modulation mode optimises the spectral efficiency and limits the interference between adjacent channels, even for high data rates. The receiver operation process must adapt to transmission changes due to multiple channel paths and co-channel interference (see Chapter 1.3).

4.3.3 Logical Channels

The physical channel of the TDMA frame is a basic time interval of 577 μs which carries a number of logical channels. The logical channels transport either user data during a call or signalling information intended for the mobile terminal or the base station. Two groups of logical channels are defined:

- traffic channels, for call data;

- signalling channels, to communicate service data between network equipment nodes.

Traffic channels
A multiframe 26 carries 24 traffic frames (Traffic Channel (TCH)); a signalling frame (Slow Associated Channel (SACCM)) and the last frame is not used. TCH transport either telephony or data, and are divided into two sectors: full-rate channels and half-rate channels. Half-rate channels are obtained by using, on average one packet burst out of two, and provide a bit rate for coded speech of 6.5 kbit/s. The use of half-rate channels allows the doubling of network capacity. A full-rate channel gives a transmission bit rate of 13 kbit/s for speech. Table 4.2 lists the channels used in the exchange of information between a base station and a mobile station.

Table 4.2: Traffic channels

	Acronym	Transmission direction	Application
Full-rate speech	TCH/FS	BS ⇔ MS	Speech
Half-rate speech	TCH/HS	BS ⇔ MS	Speech
9.6 kbit/s full rate data	TCH/F 9.6	BS ⇔ MS	Data
4.8 kbit/s full rate data	TCH/F 4.8	BS ⇔ MS	Data
2.4 kbit/s full rate data	TCH/F 2.4	BS ⇔ MS	Data
4.8 kbit/s half rate data	TCH/H 4.8	BS ⇔ MS	Data
2.4 kbit/s half rate data	TCH/H 2.4	BS ⇔ MS	Data

Signalling channels
The signalling channel group is larger than the traffic channel group, and is divided into four sectors. These are:

- broadcast channels, Broadcasting Channel (BCCH);
- common control channels;
- dedicated channels;
- associated channels.

Tables 4.3–4.6 list the broadcast channels used for dialogue between the base station and a mobile terminal in the above sectors respectively.

A traffic channel transports either speech or data. A full-rate logical traffic channel can transmit coded speech at 13 kbit/s or data at the standard rate of 300–9600 bit/s. During a call, a signalling channel is associated with a traffic channel and supports the radio link between the mobile terminal and the

Table 4.3: Broadcast channels

	Acronym	Transmission direction	Application
Broadcast channel	BCCH	BS ⇒ MS	Dissemination of general information
Synchronisation sub-channel	SCH	BS ⇒ MS	Mobile terminal synchronisation
Frequency control sub-channel	FCH	BS ⇒ MS	Mobile terminal frequency control

Table 4.4: Common control channels

	Acronym	Transmission direction	Application
Access grant channel	AGCH	BS ⇒ MS	Resource allocation
Paging channel	PCH	BS ⇒ MS	Calling a mobile station
Random access channel	RACH	BS ⇐ MS	Resource request by the mobile station

Table 4.5: Associated channels

	Acronym	Transmission direction	Application
Full rate fast associated control channel	FACCH/F	BSc ⇔ MS	
Half-rate fast associated control channel	FACCH/H	BS ⇔ MS	User network signalling
Slow associated control channel (full rate)	SACCH/TF	BS ⇔ MS	Transport of radio layer parameters: this channel is linked with the TCH/ FACCH or SDCCH
Slow channel	SACCH/TH	BS ⇔ MS	
Slow associated control channels (half rate)	SACCH/C4	BS ⇔ MS	Transport of radio layer parameters: this channel is linked with the SACCH
Slow associated channels	SDCCH/C8	BS ⇔ MS	User network signalling: this channel is linked to the SACCH

Table 4.6: Dedicated signalling channels

	Acronym	Transmission direction	Application
Dedicated control channels	DCCH	BS ⇔ MS	
Stand alone dedicated control channel: these share the same physical channel	SDCCH/4	BS ⇔ MS	User network signalling: linked to SACCH
Non-associated channels: these are in separate physical channels	SDCCH/8	BS ⇔ MS	User network signalling: linked to SACCH

base station (transmitter power control, quality of transmission measures, etc.). It carries non-urgent messages. In the call set-up and release phases, the traffic channel is temporarily used as a rapid signalling channel.

When a call is not in progress, the signalling channel fulfils the following functions:

- synchronisation of the mobile terminal, using the synchronisation channel (SCH);
- identification of the base station using the broadcast control channel BCCH;
- paging a mobile terminal, using the paging channel (PCH);
- point-point signalling, using the stand-alone dedicated control channel (SDCCH);
- subscriber access authorisation, using the access grant channel (AGCH);
- request for a dedicated channel by the mobile terminal, using the random access channel (RACH).

Terminal synchronisation

The base station periodically broadcasts dedicated timing packets on the SDCH channel in order to synchronise mobile terminal clocks.

Base station identification

The base station regularly broadcasts general information to mobile terminals on the BCCH, allowing their locations in the network to be monitored, and updating them with the options in force. This information provides them with the name of the base station, the name of the cell, the zone in which they are situated, and the signalling channel DCCH.

Terminal paging
When an incoming call arrives, the base station sends out a request on the PCH channel to mobile terminals requested by the call to activate the call establishment process.

Allocation of traffic channel
An access authorisation channel allocates a physical traffic channel to a logical connection between a mobile terminal and a base station.

Subscriber authentication
Before a traffic connection can be established, the subscriber and the terminal must be authenticated, and the signalling channel conveys the necessary information for this.

Mobile terminal request
A random access channel sends out a request to a mobile terminal activating an exchange of information.

Channel coding
Different types of interference can corrupt a radio transmission (see Chapter 1.3):

- interference;
- intermodulation distortion;
- channel interference;
- multi-path fading.

Coding of the information transmitted in the physical channel is aimed at minimising the error rate due to the different sources of radio interference mentioned above. Three methods of coding are used together:

- Block coding, with a parity bit added to a data block in order to detect odd numbers of errors in the code block.

- Convolution coding, applied to blocks on which a maximum likelihood estimation is performed using the Viterbi algorithm. With convolution coding, the data is not split into independent messages. Instead, the resulting code is a sequence of bits to which regularly spaced redundancy bits have been added. Whatever the origin chosen, in a sequence of N bits, there are M bits for the useful information and $K = N - M$ redundancy bits.

- Block interleaving in groups of 464 bits.

GSM does not use the traditional speech coding method of Pulse Code Modulation (PCM) using A-law conversion, because this implies too high a data rate (64 kbit/s) for the radio channel. As a result, GSM uses a less demanding coding law (RPE-LTP) which can operate at 13 kbit/s full rate, or 6.5 kbit/s half-rate without compromising sound quality significantly.

Figure 4.6 shows the stages of channel coding.

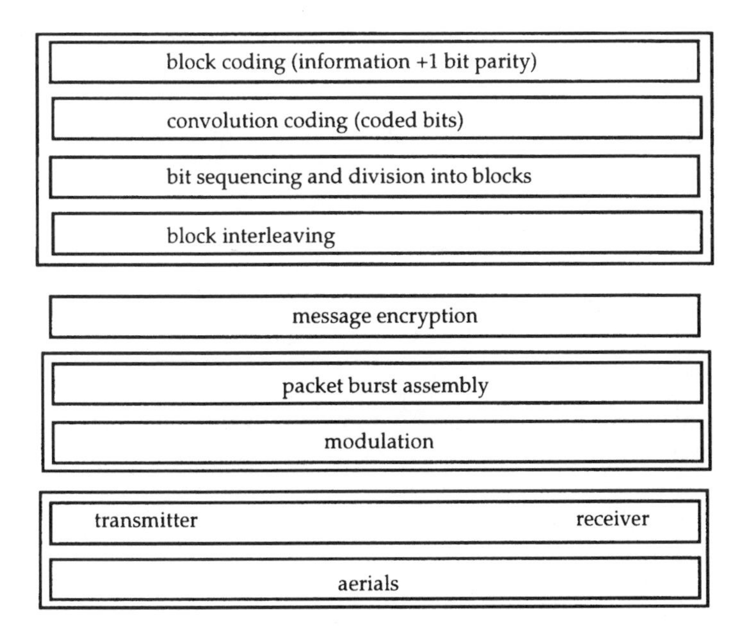

Figure 4.6: Coding of transmitted data

4.4 Layer 2: Datalink

The OSI layer 2 protocol manages the signalling between the different network entities (mobile station, BTS, BSC, MSC, VLR and HLR).

Three groups of protocols are employed in GSM layer 2:

- LAPDm – link access protocol on mobile channel D;
- LAPD – link access protocol at the A-bis interface level;
- MTP – CCITT message transfer protocol.

The LAPD and LAPDm protocols used in the radio sub-system are very similar to the ISDN protocol. However, LAPDm exploits to advantage the synchronised transactions in order to remove the need for flags, and to increase speed and protection against errors.

Telephony can be transmitted at a bit rate of 13 kbit/s in the radio subsystem, and this allows four channels to be multiplexed on one PCM time slot in the BTS ⇔ BSC link (64 kbit/s = 16 kbit/s × 4), resulting in a reduction of transmission costs. The transcoding of 13 kbit/s voice coding of GSM to 64 kbit/s using A-law coding, as in the fixed network, only takes place at the switching centre (MSC). However, this transcoding from GSM to A-law can take place anywhere after the base station, making use of standard transmission equipment deployed throughout the network.

The MTP protocol retains the full ISDN functionality.

4.5 GSM Protocol Stacks

Figure 4.7 shows the GSM protocol architecture. The base station and the base station controller are the conduits between the mobile terminal and the network sub-system:

- The Call Control application (CC) manages the call processing (set-up, supervision, clear down);
- The Short Message Services application (SMS) controls messaging;
- Supplementary Services (SS) controls the associated services;
- Mobility management controls the location of the terminal;
- Radio Resource management (RR) manages the radio link.

The service applications CC, SMS and SS originate in the terminal equipment and are carried in a transparent way through the relaying equipment BSC and BTS.

The mobility management application is situated in the network subsystem and the mobile terminal, because both of these entities need to know and store the mobile terminal location in the network.

The management of RR concerns the mobile station and the radio subsystem, and the allocation of frequencies is decided by the base station controller according to the demands of the network at the time.

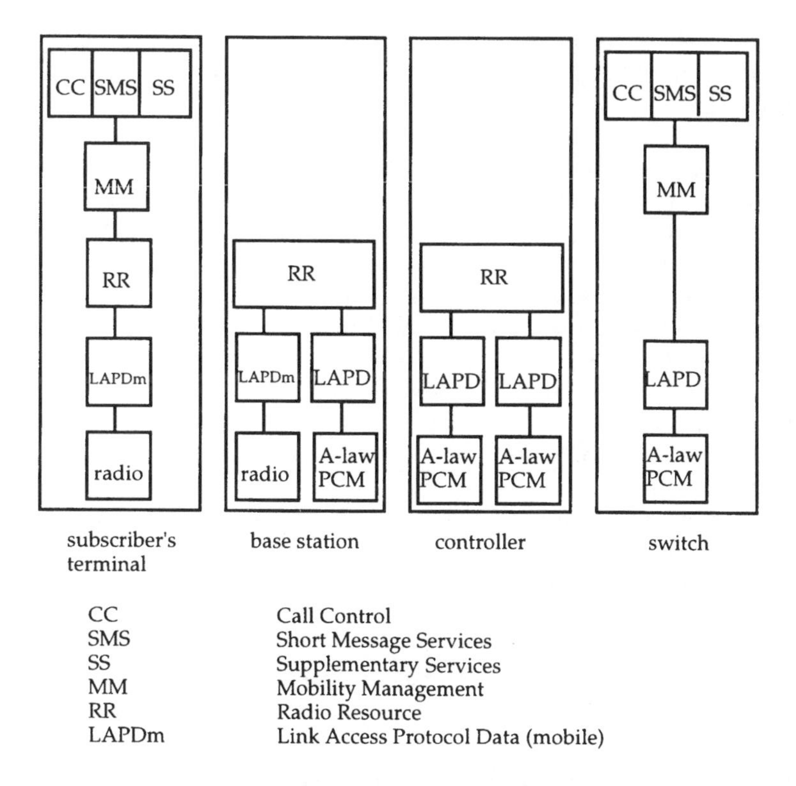

Figure 4.7: Protocol stacks for the mobile station and the network sub-system

The A-bis interface is situated between the BTS and the BSC. The physical layer operates at 2 Mbit/s PCM, and the layer 2 protocol is LAPD.

The interface A between the BSC and the network sub-system uses CCITT signalling system No. 7.

4.6 Radio Characteristics of the GSM Standard

The main characteristics are the following:

Mobile terminal to base station transmission frequency 90–915 MHz
Base station to mobile terminal transmission frequency 935–960 MHz
Frequency bands available 25 + 25 MHz
Access mode TDMA/FDMA
Radio channel spacing 200 kHz
Upstream/downstream frequency spacing 45 Mhz

Number of channels in each direction	124
Number of full-rate speech channels per radio channel	8
Type of transmission	digital
Overall bit rate of each radio channel	270 kbit/s
Overall bit rate of each full-rate telephony channel	22.8 kbit/s
Full-rate Codec bit rate	13 kbit/s
Type of speech coding	RPE-LTP
Type of modulation	GMSK
Maximum power output of a mobile terminal	8 W
Maximum power output of a hand-held mobile terminal	2 W
Maximum cell radius	30 km
Minimum cell radius	200 m
Maximum data rate	9.6 kbit/s
Automatic cell handover	yes
Roaming	yes
Subscriber identity card	yes
Authentication	yes
Radio interface encryption	yes
Transmitter power control	yes

4.7 Radio Resource Management

Layer 3 resources control the links between the mobile terminals and the infrastructure. When a terminal is switched on, it scans the radio channels, looking for the logical synchronisation channel. Once synchronised, it goes into standby mode, waiting to receive a message on the search channel, or to issue a request for access to the network on the random access channel. In the latter case, a specialised channel is allocated via the access authorisation channel.

Inter-cell channel transfer

This a major function in GSM. Before implementing a channel changeover, the BSC analyses data relating to traffic in the cells, and data about the radio links, the quality of the links (error rates), and about the received signal strength and the timing advance. It then decides upon either a channel transfer within the cell, or a changeover to a channel in an adjacent cell. The measurements of radio link quality are made by the base station and the mobile terminal. The mobile terminal measures the link quality of its current logical channel and the characteristics of adjacent cells on request from the base station (cf. adjustment of timing advance).

The BSC is aware of the criteria for initiating an intercell channel transfer, but these rules are not the subject of GSM recommendations. Different manufacturers may have different criteria, which the network operator will take into account.

Control of transmitted power

The emitted radio powers of the mobile terminal and the base station are continuously adjusted (every 60 ms) in order to limit radio interference and corruption of the data transport, and to improve the spectral efficiency; and also to increase the battery discharge time in the mobile terminal by optimising energy usage.

Adjusting the timing advance

Since the terminals present in a cell are at various distances from the base station antenna, the propagation times of their transmissions differ, and this necessitates a guard time band between the end of transmission from mobile M_n and the start of that on mobile M_{n+1}. In order to reduce this guard time, the base station continuously monitors it, and the base station controller adjusts it as a result for each of its terminals. It uses this measure as a criterion in deciding whether to perform an intercell channel handover.

Controlling the radio channels

The BSC manages the radio channels in the context of configuring physical and logical channels available by the network operator within the network plan. In this regard, GSM standards give great freedom to the operator to define its own logical channel allocation strategy.

User categories

When network access requests by users become greater than the number of available channels, the base station selects the users whose demands will be met, according to their user category. This category is defined by the network operator when subscription is taken up, and forms part of the data recorded on the subscriber SIM card. Table 4.7 lists the user categories defined in the standard.

Table 4.7: User categories

0–9	Ordinary subscriber
11	Reserved for the network operator
12	Security services personnel
13	Utilities service (water, gas, electricity, etc.)
14	Emergency services
15	Network operation personnel

The Radio Sub-System

5

5.1 Overview of the Radio Sub-System

A radio sub-system consists of one or more base stations and their associated Base Station Controller (BSC). This ensemble administers the radio channels according to the network layout. Figure 5.1 shows a schematic view of the GSM radio sub-system.

The Base Station (BTS) manages the interface between the GSM infrastructure and the mobile stations, and the BSC governs one or more base stations determined by the network architecture. This configuration in turn depends on the constraints imposed by the physical topography and the density of subscribers in the area served. The BSC controls the radio frequencies used by the different base stations, and also controls the base station OMC functions, which carry out their tasks remotely under its direction. It automatically takes charge of the inter-cell handovers of the mobile stations which are moving within its coverage area. A BSC has three standardised interfaces to the fixed network:

A-bis	with base stations;
A	with the network sub-system;
X.25	with the operations and maintenance centre.

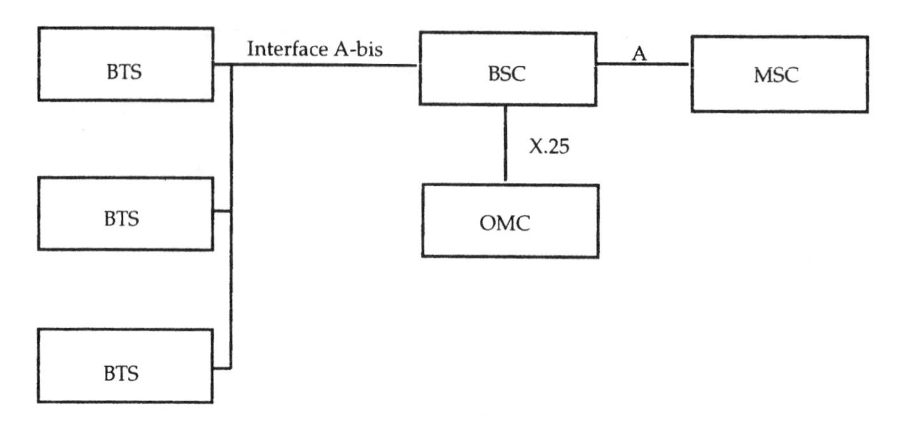

Figure 5.1: The GSM radio sub-system

5.1.1 The A-bis Interface

The physical layer is defined by a 2 Mbit/s PCM link and the datalink layer uses the LAPD protocol.

On the radio interface of a base station, a telephony channel has a bit rate of 13 kbit/s, but the PCM link has a bit rate of 64 kbit/s. In order to bridge this difference in bit rates, two options are available on the A-bis interface:

- multiplex four telephony channels onto one PCM channel;
- transcode the telephony channels up to 64 kbit/s.

The first solution has the advantage of reducing the requirements and the costs of transmission hardware between the base stations and the BSC, where the traffic is concentrated. However, the second solution has the advantage of making the transmission equipment common across the network, although the transmission capacity is not used to maximum efficiency. The multiplexing and transmitting equipment are transparent to the protocols.

Speech transcoders adapt the lower bit rate GSM data (13 kbit/s) used on the radio channels up to the 64 kbit/s used in the fixed network. The transcoders are generally installed between the BSC and the network sub-system, but to make best use of the potential of GSM speech coding, the transcoders are usually located at switch sites, although they may also be situated at BSC sites.

5.1.2 The A Interface

The physical layer is defined by a 2 Mbit/s PCM link, and the datalink layer for data is the CCITT signalling system No. 7.

Figure 5.2 shows the software layers of the A-bis and A interfaces.

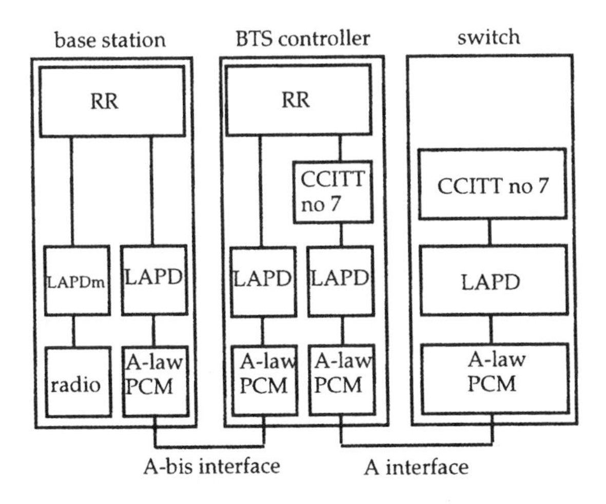

Figure 5.2: Software stacks for A-bis and A interfaces

The requirements of radio coverage dictate that the configurations of radio sub-systems listed below should be catered for. These configurations should be able to adapt to all types of zone, to all types of geographical topography, to rural areas of low traffic density and to urban areas of high traffic density. Table 5.1 lists the existing types of configuration.

Table 5.1: Configuration types

Type	Description
Omnidirectional	A BTS and a BSC share the same site
Cluster	Several BTS are linked to one BSC (in a star, ring or chain)
Sectorised BSC	Three BTS and one BSC share the same site, or the BSC is on a distant site

The omni-directional configuration in which one BTS and one BSC are together on the same site is intended for rural areas. The other configurations are more appropriate to urban sites. Figure 5.3 shows the radio sub-system configurations.

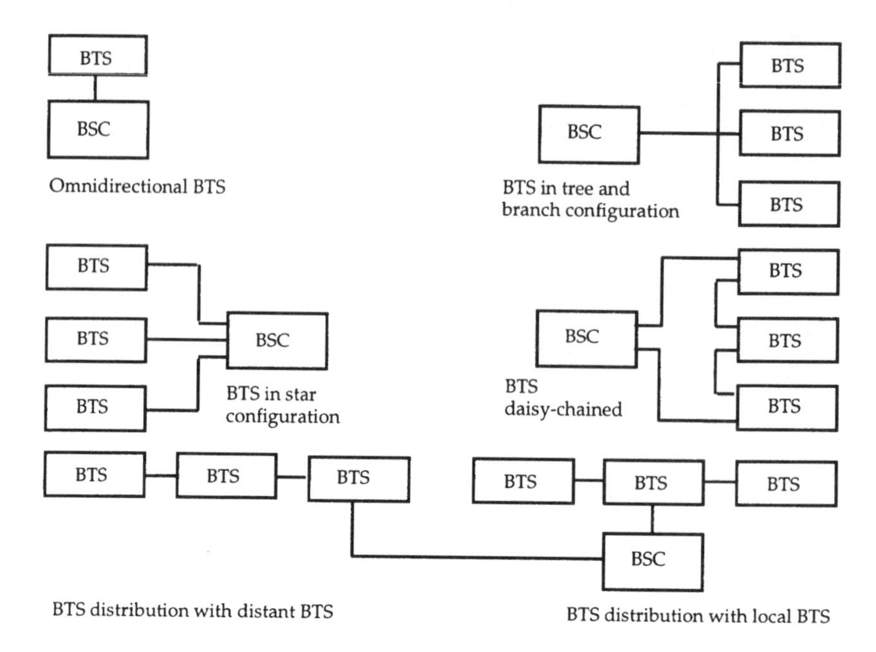

Figure 5.3: Radio sub-system configuration

5.1.3 The X.25 Interface

The X.25 interface links the BSC and the operations and maintenance centre. Figure 5.4 shows the relationships between the radio sub-system and the operations and maintenance centre of the network.

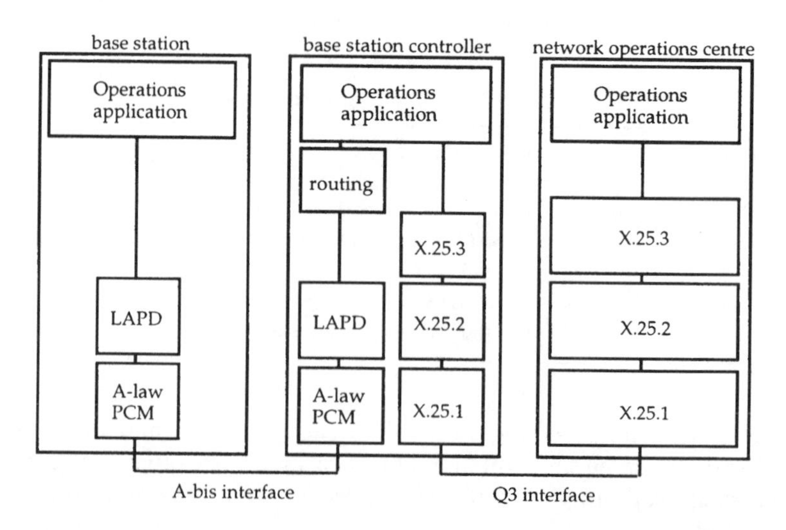

Figure 5.4: BSC to OMC interface

5.2 The Base Station

The base station is the gateway to the network for the mobile stations. To achieve this, it performs the role of radio infrastructure hub. Mobile terminals are linked to the base station through the radio interface (Um) and communications with the BSC are conducted through the A-bis interface. A base station controls from one to eight radio carriers, and each radio carrier provides eight full-rate radio channels, transmitting via an omnidirectional or a directional antenna (usually 120° sector). In the cases of sectorised coverage, a single site can accommodate several base stations, which are synchronised in order to increase the efficiency of inter-cell handovers.

The base station functions are:

- radio transmission in the GSM format, employing frequency-hopping techniques and spatially diverse antennas;
- implementation of equalisation algorithms to counter the effects of multiple paths;
- coding and decoding of radio channels;
- encryption of transmission data streams;
- control of the protocols governing messaging in the radio datalink layer (LAPDm);
- measurement of quality and received power on the traffic channels;
- transmission of signalling messages;
- operations and maintenance of the BTS equipment.

To meet these needs, the functional architecture of the base station is as shown in Figure 5.5.

The equipment making up a base station includes:

- a timing reference;
- a management unit;
- a frequency-hopping circuit;
- a radio transceiver unit;

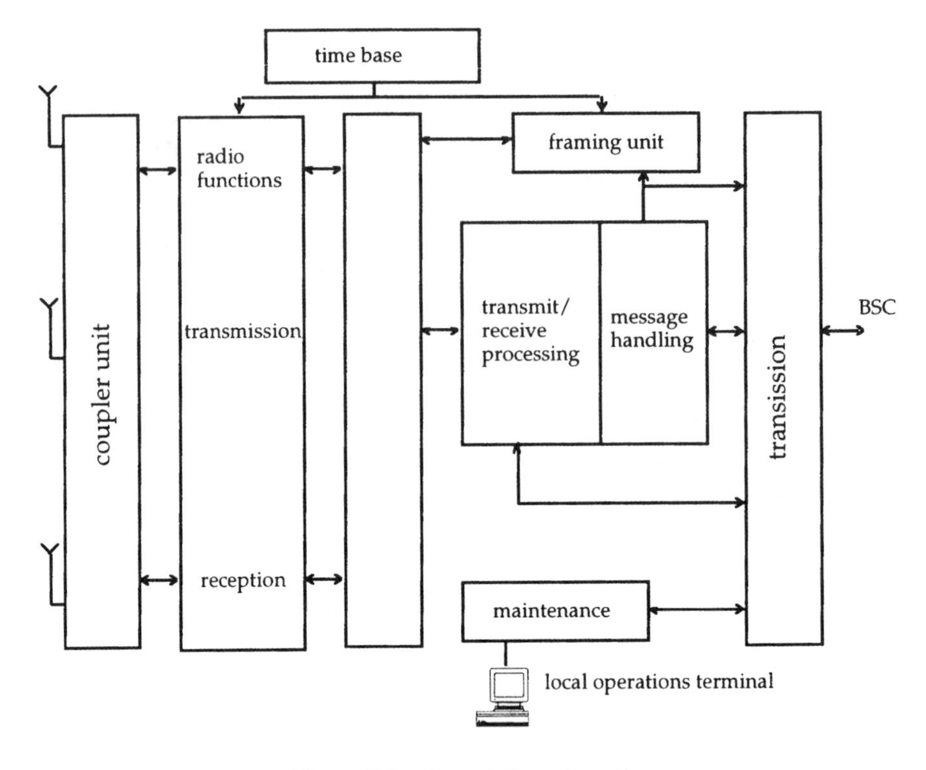

Figure 5.5: Base station schematic

- a frame generator;
- a radio feed unit.

Timing reference unit
The timing reference provides all the synchronised clock signals needed by the other circuits in the station to define the time reference intervals specified in the GSM synchronisation standard. These are:

- a quarter bit (0.92 μs);
- a time slot interval (577 μs);
- a TDMA frame time (4.615 ms);
- a multiframe period;
- a superframe period;
- a hyperframe period.

Maintenance unit
The maintenance unit fulfils the following roles:

- Management of all internal communication protocols and acceptance of alarms from all equipment in the BTS;
- filtering and relaying of alarms to the BTS;
- forwarding of commands from the BSC;
- input and updating of software and files controlling unit configuration and operation;
- control of the man–machine interface for local management of the station.

Frequency-hopping unit
This unit brings about the switching between the frame generators (base-band section of the BTS) and the carrier frequency controller unit according to a given frequency-hopping algorithm in order to trigger a frequency jump at each time slot.

Radio monitoring unit
The radio monitoring circuit detects and localises faults in the transmitter-receiver chains. To aid this process, it allows looping of the radio circuits.

Frame generator
The frame generator contains all the necessary base-band data processing functionality needed to cater for eight full-rate channels, or 16 half-rate channels. It controls the layer two signalling LAPDm and LAPD protocols for communication with the mobile terminal and the BSC, respectively. The frame generator unit governs the radio channels, adjusts the transmitter power and controls the radio transmission quality. The pre-transmission processing functions are: the alignment of the data bit rate and the speech bit rate; channel coding; interleaving; encryption; and building the frames. At the receiver, the reverse of these operations is performed with, in addition, demodulation, equalisation and error rate measurement being carried out.

Radio transceiver unit
This unit contains the base station transmitter and receiver. The transmitter modulates, performs shifts in radio frequency and amplifies the power. The receiver performs the inverse frequency shift, the analogue to digital conversion, and calculates the signal strength.

Radio coupling unit
The radio coupling unit includes multiple couplers to facilitate reception diversity (two antennae are coupled into the receiver circuitry), and cavity transmitter filters, which minimise the power coupling loss. These filters are tuned remotely by the network operations centre, in an operational function which allows the selection of base station carrier frequencies to made at the OMC.

Traffic and expansion unit
This is installed in accordance with the traffic density to be handled by a cell. Expansion units are baseband processing units or frame units in the radio frequency media (carrier and coupling units).

Transmission equipment
The transmission equipment governs the interface to the BSC. It allows up to 80 full-rate radio transmission channels to be multiplexed on to one 2 Mbit/SPCM link. In urban areas, a base station controls three cells. Such a station may handle eight carrier frequencies per cell, or a total of 24 carriers, equivalent to 192 (24 × 8) radio communication channels. A BTS builds in a great deal of flexibility, and all software can be downloaded remotely. Interface standardisation allows the network operator to build a network from equipment procured from a range of suppliers.

A typical base station has between one and four transmitter–receivers, is powered from the mains, and has a back-up battery. The availability of highly integrated ASIC chips, which are custom designed for the base station electronics, has led to a reduction in the size of base stations, which can now be installed in confined locations.

5.3 The Base Station Controller (BSC)

The BSC controls the radio sub-system, and its main function is to oversee the base stations. The controller can be co-located with the base station, sited in the switching centre (MSC), or located on an independent site. In the latter case, it carries out the role of concentrating the base station traffic, which optimises the network transmission. Figure 5.3 shows the possible BTS–BSC configurations.

The functions of a BSC are:

- management of radio resources (traffic channels, signalling channels, etc.);

- call control (set-up, supervision, and clear-down);
- management of inter-cell handover;
- control of transmitted power;
- operations and maintenance management, and signalling to the OMC;
- management of security and reconfiguration processes;
- management of alarms and supervision of peripheral equipment;
- protection of software and base station configuration data.

In order to achieve high reliability and equipment availability, the key hardware elements are duplicated.

Physical architecture
A BSC operates around a switching matrix. This allows traffic routing from one input point to several destinations. Figure 5.6 shows the BSC physical architecture.

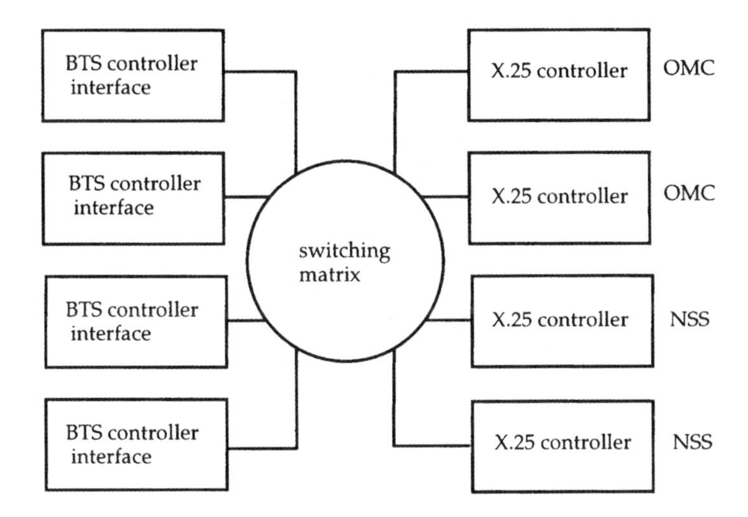

Figure 5.6: BSC structure

The controller acts as a communication cross-roads, and it is shown in the centre since it acts as an intermediary to all of the inputs. Three types of controller card govern the different interfaces: communications with the BTS, communications with the OMC, and communications with the network sub-system (NSS).

Switching matrix

The switching matrix has 64 input ports, each of which can switch a PCM time slot from one port to another. As well as switching 64 kbit/s circuits, the matrix transfers operational data between the BSC controller cards. Several controllers can be linked by an access bus in the matrix. By this means, the available PCM circuits are shared between the controllers on the bus.

Controller X.25 interface

The X.25 controller interface terminates the link to the OMC. It implements maintenance functionality, and the memory that holds the software of the entire radio sub-system is protected. This controller card also manages the radio resources. Figure 5.7 shows the controller interface software architecture.

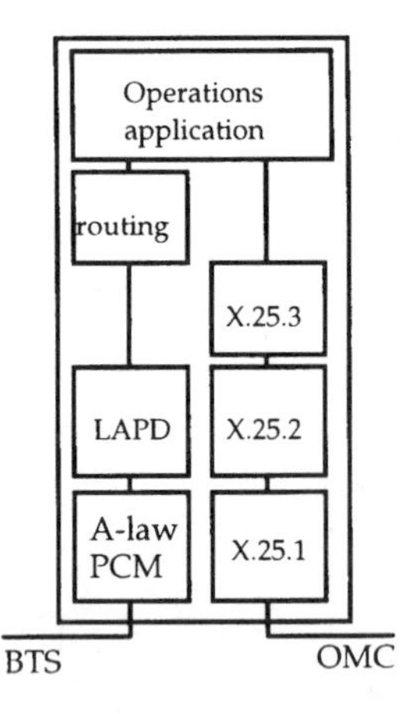

Figure 5.7: Software architecture of the X.25 interface

Controller interface serving the BSC–BTS interface

The interface controller serving the BTS controls the datalink layer LAPD signalling protocol. It also controls the radio carrier (eight physical channels) allocated to the BTS, the BTS signalling, and the quality of service measurements on the channel traffic.

The NSS interface controller

The NSS interface controller terminates the 2 Mbit/s PCM link to a network switch, and controls the CCITT No. 7 signalling and the HDLC physical layer protocol.

The Network Sub-System

6

6.1 Overview of the Network Sub-System

The network sub-system is the bridge between the radio section of the GSM network and the public switched telephony and ISDN networks. It is complex, and Figure 6.1 shows the main integrated equipment elements, and the links between them.

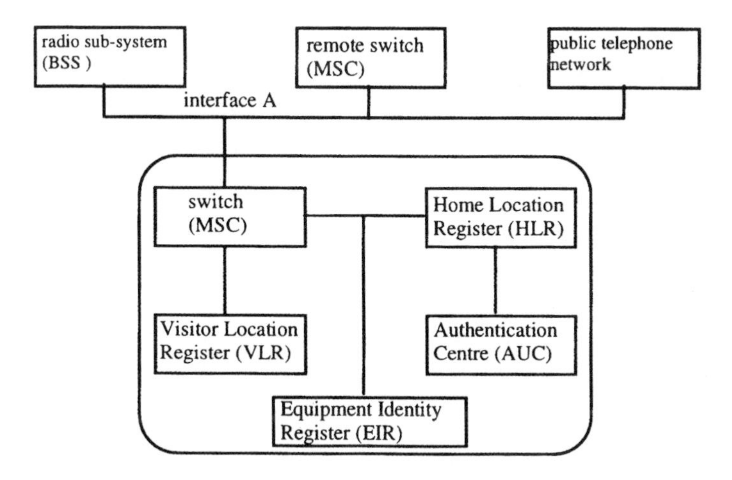

Figure 6.1: The GSM network sub-system

6.2 The Mobile Switching Centre

The Mobile Switching Centre (MSC) is the most important of the sub-system elements, being responsible for the switching functions necessary to interconnect mobile subscribers, and for connecting mobile subscribers to the fixed network. The switch interfaces to the public switched telephony and ISDN networks and to the public data packet-switched and circuit-switched networks. The MSC contains three sorts of database, in which it stores and retrieves information on call processing, and information to meet subscriber requests for service. These are:

- the Home Location Register (HLR);
- the Visitor Location Register (VLR);
- the Authentication Centre (AUC).

Each of the MSCs takes part in the updating of the data bases with the latest information available to it, and this gives rise to frequent message dialogues between the switches. Three types of services are provided to subscribers by the switch:

- support services – 3.1 kHz audio, synchronous data transmission, packet assembly/dis-assembly and switching between data and telephony modes;
- teleservices – telephony, emergency calls, fax;
- supplementary services – call diversion, charging advice, call barring.

6.3 The HLR

The HLR stores information relating to certain mobile subscribers, and this forms part of the network reference data. Each mobile subscriber has a register of fixed data describing the type of subscription account and the services that the customer has subscribed to, and a register of temporary data describing the subscriber's current location and the status of his terminal. The subscriber's HLR remains fixed throughout the duration of his subscription. It is a protected database, and the encrypted data is only handled by a single operator: in addition, the HLR is sited in a building with restricted access.

Incoming calls for a mobile subscriber can be processed by reference to the location information held for the called subscriber. For this, the network finds the mobile's last known location and its status, and, if the terminal is free, it interrogates the nearest location register to check whether this location is correct. The switch initiating the call is then advised of the position of the mobile terminal, following which it interacts with the switch nearest to the called subscriber in order to set up the call. Figure 6.2 illustrates this process.

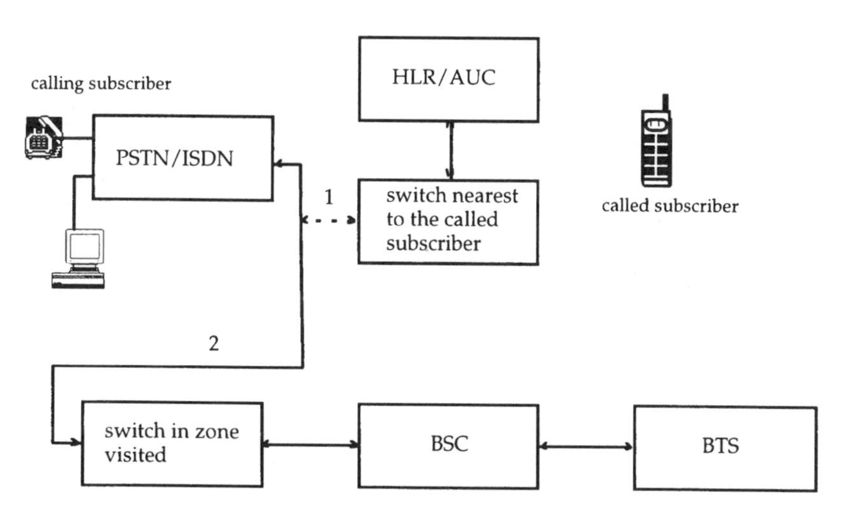

Figure 6.2: An incoming call for a mobile terminal

The network first interrogates the subscriber HLR since it knows only the member of the called subscriber; this is route No. 1. Following this, the network accesses the VLR in the zone in which the called subscriber is currently situated (route No. 2), and the call is finally routed through to the subscriber via the BSC and the BTS of the relevant cell. The call could be initiated from an ordinary telephone or from an ISDN data terminal.

6.4 The VLR

The VLR records data relating to the mobile subscriber when he enters the coverage zone of the network sub-system. The VLR is a dynamic database, and it communicates with the HLR of a subscriber in order to obtain the information relevant to the processing that it will be requested to perform on behalf of the subscriber. The subscriber data is transmitted to another VLR when the subscriber moves out of the zone covered by the MSC, and this data generally follows the subscriber as he moves significant distances through the network. The subscriber data allows his incoming and outgoing calls within the coverage zone of an MSC to be processed.

6.5 The AUC

The role of the AUC is to police the identity of network users and mobile stations; in other words, to protect the network against potential intruders. The AUC stores all the data necessary to protect mobile calls. Two key provisions of the GSM standard are the encryption of radio channel transmission and the authentication of network users. The encryption keys are held in both the mobile station and the AUC. This information is, of course, protected against all non-authorised access. Protection measures are also taken when the subscriber is entered on to the system and the network gives the subscriber his personal secret key, Kp. This key is encrypted using a separate algorithm, and is then stored in the database in encrypted form to prevent it being read directly from the memory storage medium.

The radio link between the terminal and the fixed network infrastructure can also be the subject of fraudulent call attempts from users. Three levels of protection have been put in place to counter this danger:

- Subscriber Identity Modules (SIM) are authenticated by the system to prevent the network being used by non-registered users;
- radio channel communications are encrypted to prevent eavesdropping on either telephony or data calls;
- a subscriber's identity is protected.

Security is achieved by the use of encryption algorithms in the terminals and in the infrastructure. Administrations belonging to CEPT have defined a set of algorithms for this purpose, but other algorithms may be defined by other types of network operator. The operator may choose to enable or disable the encryption mechanisms in the radio links according to the priorities. The authentication data and the encryption keys are kept in the AUC.

The options set up by the standardisation committee allow network operators to choose their own encryption algorithm independently of other operators, which allows them to authenticate visiting users without knowing the algorithm used in their home networks.

The authentication of a subscriber is arrived at by requesting his terminal to submit the result of a calculation performed on a random number supplied by the system, by applying his personal key, Kp, which is stored on his SIM. The system compares the result output by the terminal with the value expected. This calculation is performed according to a set algorithm and is based on a secret key which is unique to the SIM. The algorithm and the secret key are stored in protected format on the SIM and in the HLR. The encryption of the radio burst packet uses a second algorithm applied to another key (Kc), selected when the subscriber is connected to the network, and a further number, which is changed for each packet burst. The key Kc is selected by the terminal and the HLR with the aid of a third algorithm. Figure 6.3 illustrates schematically the subscriber authentication mechanism used in the network.

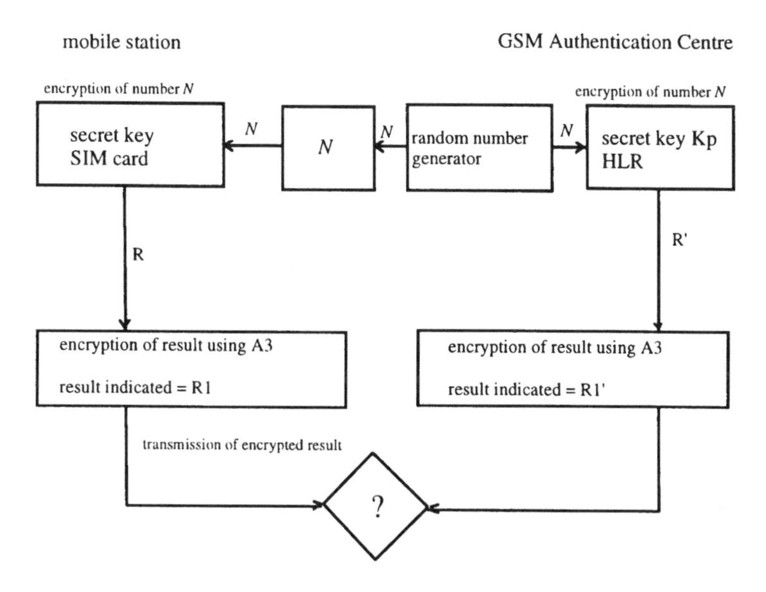

Figure 6.3: Subscriber authentication

A subscriber's personal identification number (International Mobile Subscriber Identity (IMSI)) is only disclosed on the network when the terminal makes contact with the network. After this, the identity of the subscriber is protected by the use of a substitute identity number (Temporary Mobile Subscriber Identity (TMSI)) allocated by the network when the mobile terminal enters any given zone. The terminal presents itself to the network by declaring its IMSI in the zone in which it is situated. From the IMSI, the network learns the personal subscriber number, the name of its home network and code of the country in which its subscription is based. Its location details are recorded in the VLR and in the HLR assigned to the subscriber. Figure 6.4 shows the encryption arrangements for frames passing between the mobile station and the base station.

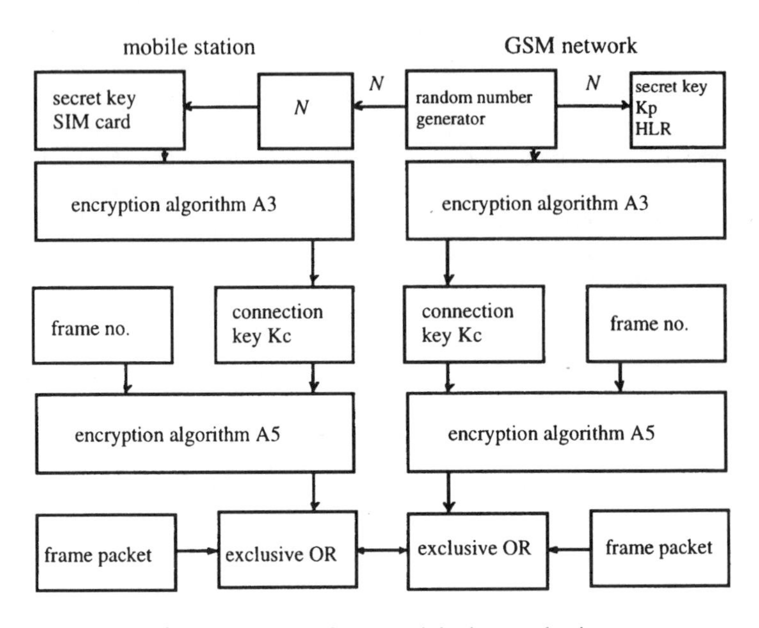

Figure 6.4: The frame enciphering mechanism

When a mobile station connects to the network, a random number generator supplies it with a number *N*, which is then encrypted with a secret personal subscriber key, Kp. The resulting number is further encrypted by applying an algorithm A3 in order to protect the key Kp. This new result is transmitted to the GSM network, which compares it with the result that it had calculated. If the values concur, the network requests the terminal to protect frames with a new key, Kc. The frames coded using key Kc are themselves encrypted using an algorithm A5. It is the product of this final encryption that is transmitted on the network. This very complex mechanism guarantees that no data is made accessible on the network in its true form. If it is now remembered that each frame is transmitted on a different frequency, it will be appreciated that the protection measures in force are numerous and sophisticated.

6.6 The Equipment Identity Registers

GSM specifications (Rec. 02.16 and 02.17) contain functions intended to discourage theft of terminals and to protect the network against the use of non-authorised terminals. These specifications affect the international identity of subscribers and the identity of GSM terminals.

Each mobile terminal has its own personal identification number, which is its International Mobile Equipment Identity (IMEI), and which has nothing to do with the subscriber identity. This identifier is installed in the terminal when it is manufactured, and it proves that the terminal conforms to GSM standards. The network checks this identity number for each call that the terminal makes, or when its location is updated in a register. If the number is not on the approved list of authorised equipment known to the network, access will be denied to it. A database located in the switch, the Equipment Identity Register (EIR), contains the list of identity numbers of certified terminals and of stolen terminals, and allows the IMEI numbers to be verified.

The switches interrogate this database to verify the validity of the IMEI of a new call, and the EIR responds to a request with a colour:

- white – for authorised equipment;
- grey – for equipment under observation;
- black – for barred equipment.

When the IMEI number appears on several lists, the indicated colour obeys a rule of decreasing priorities: black, grey, white. Traffic counters are associated with the lists indicating the number of entries and interrogations. In view of the administration burden of managing the IMEI lists, the EIR registers are only present on a few sites.

Network Management

7

7.1 Overview of Network Management

Figure 7.1 shows the relationship between the services and the equipment that supports them.

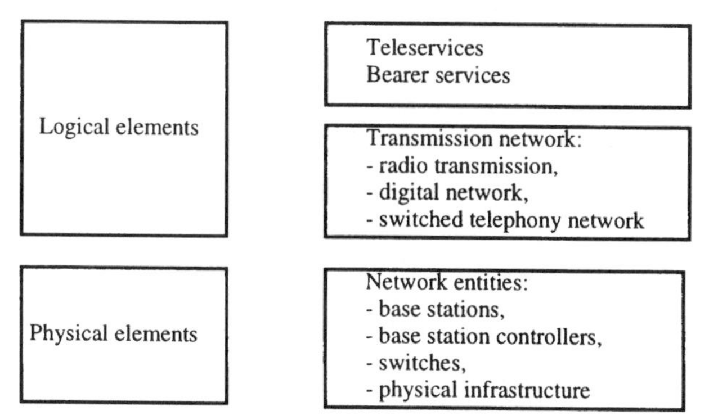

Figure 7.1: The physical layer stack and the logical stacks for equipment and services

A telecommunications network consists of not just physical equipment, but also the associated software to carry out the basic logical functions. The logical function elements serve to construct the layers of the transmission network, which are transmission 'products' (analogue or digital).

These products provide network services (bearer services, and teleservices).

The structure can be modelled at three logical levels:

- network elements (equipment and software);
- transmission network (radio, digital and switched);
- services (teleservices, bearer services).

The physical layer
This layer includes all of the physical equipment and the associated software. A piece of network equipment together with its associated software constitutes a network entity, which carries out one or more logical functions.

Transmission network layer
The network is the collection of standardised entities which are network elements (BTS, BSC, BSS, MSC and NSS), and it provides a vehicle for information transport. The transmission network is built up from logical elements formed by network entities.

Service layer
The service layer provides services with point-to-point permanent or switched links, as well as a circuit-based or packet-based transmission. These services are the products that the operator sells to customers, either as bearer services, i.e. connectivity, transmission capacity, or as teleservices, which are 'finished' products, such as telephony, fax, videotex and messaging.

In order to manage the physical and logical elements of a network, the CCITT has defined the concept of the Telecommunications Management Network (TMN). The TMN provides a modular framework in which network operators, application software and equipment communicate in a standardised and secure environment.

Network management includes the following functions:

- administration management;
- development management;
- network operations;
- maintenance;

- security management;
- evolution management.

Figure 7.2 presents a schematic view of a TMN.

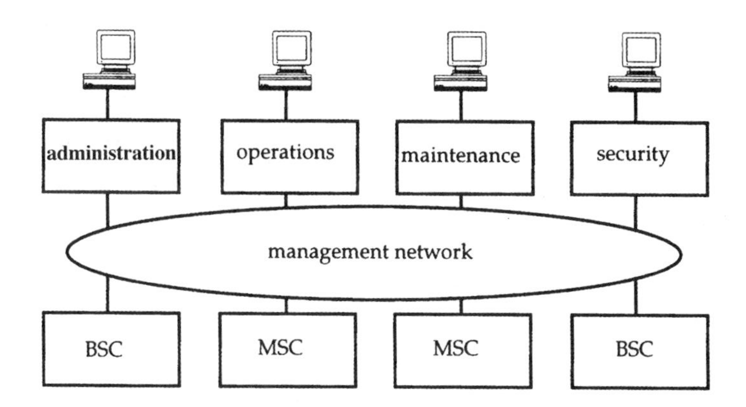

Figure 7.2: A Telecommunications Management Network structure

The telecommunications network is the logical entity providing service to subscribers, and the administration functions are carried out by management applications. The data transmission network carrying management traffic is independent of the telecommunication network itself, but interfaces with it at different points (in the physical equipment) in order to exchange information, and to control the operation of network equipment elements and logical network entities. The management network serves to control network equipment in the overall management process.

7.2 Objectives

Network management concepts are based on the protocols of (logical) dialogue and on physical interfaces defined by international bodies. The aim is to allow hardware elements made by different manufacturers to be interconnected, and to control them through standardised interfaces. Equipment is typically provided with software to enable it to function and to oversee its operational state, and also to provide integrated communications channels conveying operational data to a management centre.

The CCITT has developed the TMN concept in recommendation M.30, laying down the framework for the technical and administrative operational management of the networks. The interfaces between the equipment and the operations centre conform to the Q3 interface defined by CCITT, and this is depicted in Figure 7.3.

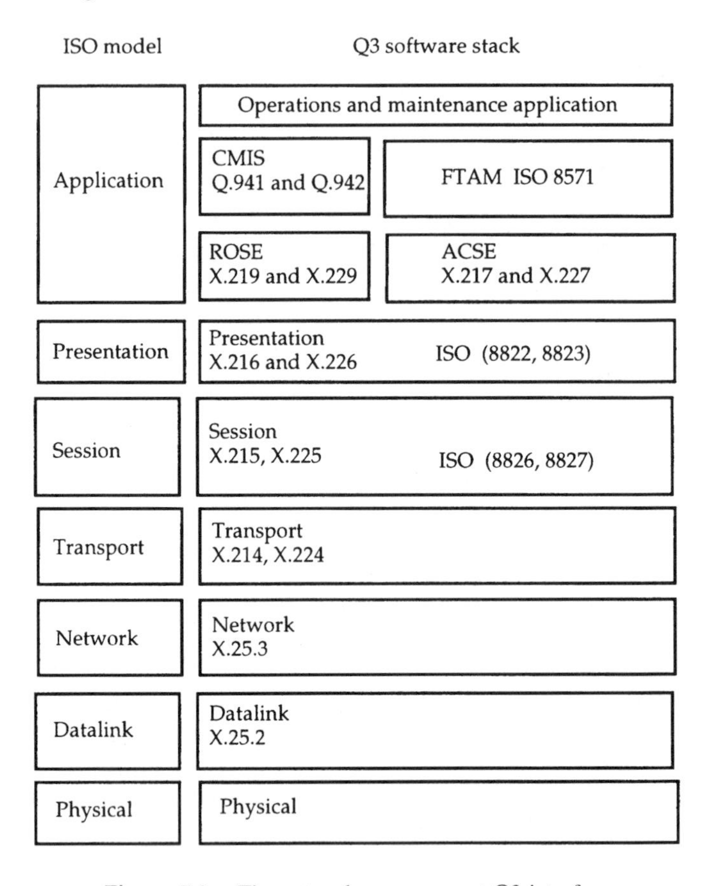

Figure 7.3: The network management Q3 interface

7.2.1 Profile of the Q3 Interface

The Q3 interface can be described using the OSI seven-layer model. The lower layers (one, two and three), which transport the management data, operate over an X.25 interface. In the application layer, there are OSI protocols providing CMIS/CMIP management services. These protocols are designed for use in an object-oriented environment, in which managed objects are described by their attributes and properties. The physical and logical network entities, their attributes and their behaviour are modelled in the management centres as logical objects. Equipment management applications run both in functional units within the operations centre (billing, network operations and maintenance), and in the equipment itself. In general, a functional management application interrogates a piece of equipment to ascertain its status, and the values of various parameters, and then processes this information. In other cases, for example in security management, an application or function within an equipment outputs an alarm flag to indicate mis-operation or the crossing of some threshold. The physical and management units incorporate the following functional entities in order to carry out management operations:

- a dialogue process that governs the logical interface for the exchange of messages;
- a database describing the set of managed objects and their associated parameters, as well as the allowed operations on these objects;
- a management application (a management agent within the physical network entities, and an element manager in the network operations entity), which establishes an association between the data and the dialogue process.

Figure 7.4 shows a simplified representation of this dialogue between functional entities.

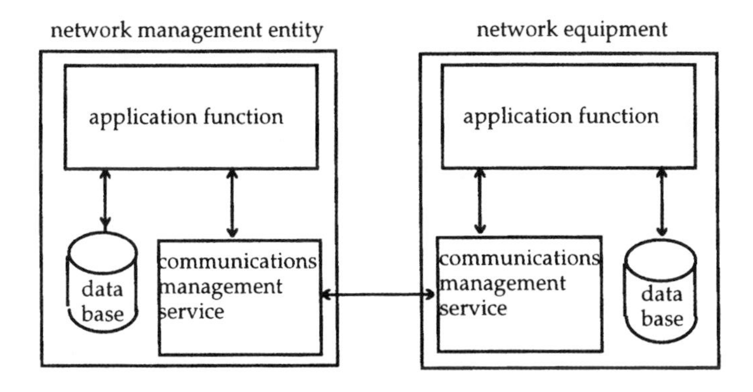

Figure 7.4: A dialogue between two functional entities

The 'communications management' function within a management application relies upon the transport layer and on the technology used. The application controls the data stored in the database, and this gives it a logical view of the network. The communications management function enables the interrogation and modification of stored parameter values to take place, and the downloading of operating software from the operations centre to the managed equipment. Management applications exist that allow the precise remote operational management of equipment. For example, the network operator can check the compatibility of the software version loaded in an equipment, or can scan the equipment event log retrospectively to establish whether any notable incidents have taken place. This type of information is useful in order to monitor the operating cycle of equipment.

Network management has two principal dimensions: (a) element management and (b) administration management. The element management domain is concerned with the following: equipment configuration (activate, shut-down, initialise, software download, start of measurement sequence, etc.); alarm management; monitoring of operating modes; and performance and updating of billing data. Element management is implemented according to the network operator's own policy. An item of equipment that receives a remote command performs certain checks on its configuration data before executing the command in order to:

- check the ability of the operator to control it;
- check the compatibility of parameters (whether they are in the correct data fields), the appropriateness of the command (whether it is compatible with the status of the equipment-out of service, in service, under test or in use);
- a check that the sequence in progress is complete and unambiguous.

Administration management includes the equipment inventory, which monitors all the system components, the suppliers, the deployment records (name, location and status) and subscriber management (the setting up of accounts, billing and commercial information).

7.3 Management Functions

Figure 7.5 shows a management structure for a GSM network.

Each of the network elements (BTS, BSC, MSC/VLR and HLR/AUC) can be locally operated from a dedicated console on the site, but centralised management is carried out from the operations centre at the OMC site. The Network Management Centre (NMC) runs the general technical administration of the network, and the commercial support operates behind this.

The 12.20 series of standards in the GSM recommendations define this structure, as well as the base elements and the network management interfaces.

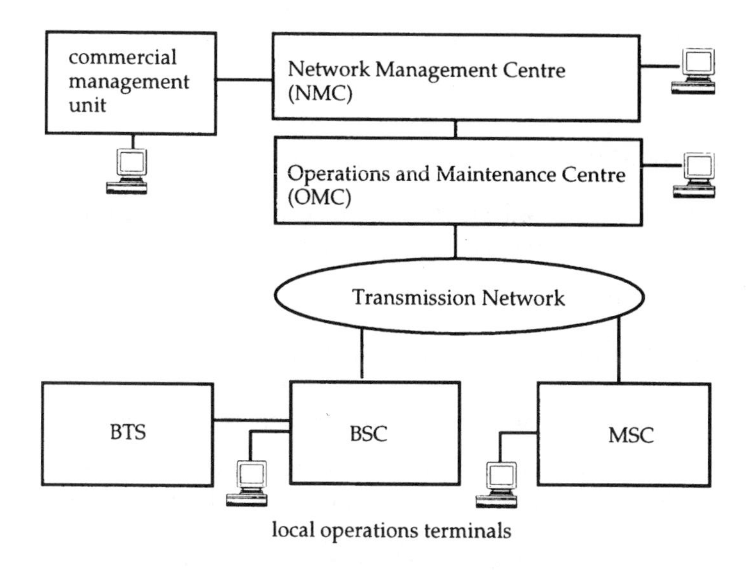

Figure 7.5: GSM management structure

7.3.1 Administration Management

Administration and commercial network management takes care of:

- the inventory of network components, with all the relevant component data (name of supplier, suppler contact name, maintenance contract, hardware version, software version, etc.);

- network component mapping, providing their location in the network;

- subscriber management (creation, modification and billing);

- database management – EIR (Equipment Identity Register) and IMEI (International Mobile Equipment Identifier), which identify each terminal;

- management of the lists of terminal status categories, which includes the following:

white list	authorised terminals
grey list	terminals put under observation by the operator
black list	barred or stolen terminals

- processing of statistics in order to plan network provisioning, traffic measurement, etc.;

- management of incident reports, fault reports and network changes, which impinge on business organisation.

7.3.2 Security Management

Network security is based on very sophisticated mechanisms. The GSM standard is committed to ensuring the following capabilities:

- subscriber confidentiality;
- authentication of subscriber and mobile identity;
- security of data and signalling conveyed across the radio interface.

Confidentiality of user identity and data in network transport is important for the following reasons: it respects the private lives of subscribers, it prevents them from being tracked, and it protects the network from intrusion or illegal access attempts by means of masquerading. Subscriber authentication is aimed not only at barring non-authorised users or terminals from using the network, but also at preventing arguments about network usage and about billing arising out of service use. The network also guarantees the integrity of the data conveyed.

Network operators define their own measures for the security of access to the databases of sensitive information (HLR, VLR and AUC). These databases are normally located on sites having access control. In addition, the data is recorded in encrypted form to guarantee its protection, even in the physical memory medium.

7.3.3 Operations

The operation and supervision of the network require administrative functions that facilitate:

- the management of the network equipment descriptive data;
- the management of equipment software for geographically dispersed equipment;
- the supervision of network performance;
- the retrospective analysis of network events;
- measurement of the quality of service given to customers.

These management activities are generally orchestrated from an operations centre, but can also run either directly on the equipment, which is provided with an interface for this purpose, or from a site having a management terminal. Management interface terminals are of two types:

- powerful terminals with graphical, user-friendly man—machine interfaces capable of operating all the operations functions;
- simple text terminals allowing description and supervision databases to be interrogated.

The objectives of network operations are:

- to guarantee subscribers an agreed quality of service through the supervision of all of the equipment (this requires direct or inferred measurement of the end-to-end quality of service);
- to supervise the equipment.

This function necessitates the collection of reports of anomalies systematically flagged by the equipment (equipment suffering faults generates visual or audible alarms for specific events, or when certain thresholds have been crossed-all of these events are archived in logs, and can be scanned retrospectively by an operations function), the analysis of the causes, diagnosis of the reasons, and the specification of repair actions or 'work-around' solutions:

- to minimise the operating costs;
- to ensure the technical management of equipment from diverse sources;
- the management of subscriber mobility-location of mobile stations, updating of the location registers, inter-cell handover;
- management of radio resources-protection against traffic overload, establishment of a connection without setting up a normal channel (the traffic channel is allocated only when the called station has answered, and a signalling channel is used prior to this), queuing of the radio resource when there is no traffic channel available when a call attempt is made;
- to maximise the productivity and flexibility of the network.

Network operations is about the optimal utilisation of the network resources in a global and compatible way, not element-by-element. The transport of data between the different operational entities (the operation of equipment, networks and services) should be optimised so that each functional area has the necessary information at its disposal. In the event of an equipment fault, it is necessary to signal the event to those able to be involved so that each one can carry out the necessary corrective actions.

All network equipment manufacturers offer their own network operations platform, usually running on a Unix-based computer. These platforms all have a user-friendly graphics interface and offer the following functionality:

- The automatic monitoring of the network topology. This is a facility that allows a zone to be displayed, and the locations of the terminals in the zone to be shown. This viewpoint allows network equipment, links between sites, and equipment and link status to be separately highlighted. There is often an option to display equipment details present on a site, as well as the status of components and external links.

- Alarm and fault management. This is the management of events having an impact on network function and, for this eventuality, alarms have an associated priority rating (1-3), describing their capacity to affect network operations. Certain equipment is duplicated, or contains redundant components in order to keep an equipment element functioning in the face of simple faults.

- The computer platform always has a statistics processing function. This allows statistics relating to alarms, traffic, faults and other significant events to be analysed. The statistical data allow the life-cycle and evolution of the network to be evaluated, and the strong and weak points- and therefore their capacity to fulfil their network functions- to be identified.

The measurement and analysis of network performance
A network operator is concerned with the performance of the network, while a user is concerned with the quality of service, based on criteria such as the ease of access, the error rate, and the average time to transfer a file across the network. However, the performance of the network often determines the quality of service, and the operator should therefore have an accurate and precise picture of his network. Performance analysis requires the prior establishment of a reference, the definition of relevant parameters and network data measurement points. Table 7.1 gives a list of parameters characterising the performance of a network. All of the main network equipment and systems manufacturers have in-house network solutions in their portfolios, often based on open Unix-based platforms. These are usually core systems offering a basic service which accommodate, through their interfaces, applications developed by service development companies. If a proprietary system is set up to closely supervise in-house equipment products, it is usually much less powerful when working with other proprietary equipment.

Table 7.1: Quality of service parameters

Quality of service parameters
Connection set-up time
Successful call set-up ratio
Speed of data transfer
Usable channel data rate
Error rate
Probability of service break due to a network fault
Call release time
Network availability

In order to measure the performance of the network, the operator must resolve problems that may arise out of different makes of equipment, where the information that can be accessed from different network components may depend on the component manufacturer. It is essential that the network operator should model his network, select the performance criteria, and select the point in the network from which data will be collected.

Figure 7.6 shows the outline software architecture of a network performance management tool.

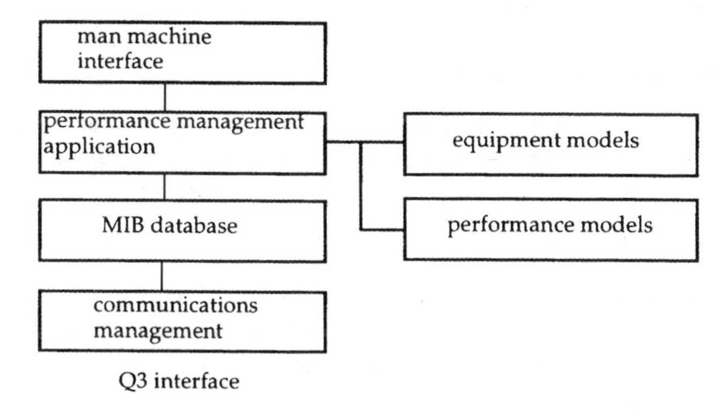

Figure 7.6: Performance management application architecture

Man-machine interface

The man-machine interface is a tool that facilitates a dialogue between one or more network operating engineers and the performance management system. It provides keyboard command input mechanisms (mouse or microphone), data output devices, a graphics interface, a printer, a loudspeaker and supporting software.

Performance model

The performance model is a reference for evaluating parameters characterising network functionality and the way in which performance varies according to the load on the network. The model aids network optimisation, which amounts to calculating the values to be allotted to certain equipment parameters from which to optimise the operation of the network. The model serves to simulate the network degradation modes in the event of faults, and therefore to evaluate scenarios which counter these fault circumstances if they actually occur, and to predict the performance parameters that will then apply.

The network operator develops and refines the performance model by building on the performance parameters indicated by equipment manufacturers and by running simulations, using analysed data obtained from network measurement exercises.

Equipment models
An equipment model is a logical description for a particular equipment type – a generic model of physical equipment and of services provided by the network, with the ranges of possible parameter values. The model allows the full description and creation of an inventory of the network components.

Management Information Base (MIB)
The MIB is the database for data describing the overall system. The MIB records three types of information:

- Static. The configuration of the network, the configuration of each of the equipment items (default values for a family of equipment), the modelling of equipment, a list of possible states, ranges of parameter values, network equipment addresses (physical and logical), etc.

- Dynamic. Equipment status (absent, out of service, in service or faulty), the alarms, responses to operator commands, and information regarding the exceeding of thresholds, clock stability in equipment, change of status messages, etc.

- Statistics. The number of packets received during an observation period, the error rate, the number of seconds of degradation, the number of successful connections, the number of connections refused, mail-box occupancy in messaging, etc.

All of this data can be used in a reboot of the network when starting 'from cold'.

The application of Performance Management
The performance management application is an OSI 'manager', which is supported by two logical reference models (a model of each of the physi-

cal and logical network components, and a model of the performance of each network component) and provides an interface through which an operator can create and modify these models, which are logical objects. The application is also an interface through which to manipulate the MIB data, and to interrogate and pass commands to the equipment. The 'manager' supervises and analyses the functioning of the network (the equipment status and the services). The application detects the variations in network performance (an item of equipment, a fault or traffic overload). It is equipped with contingency plans to direct the network in the event of a serious network failure in order to ensure that it continues to operate, albeit in a degraded performance mode, or to apply countermeasures in the case of unpredicted traffic overload which would seriously impair network performance. Rules for accepting calls in critical situations are defined after certain thresholds have been crossed, depending on the source and destination of the call, and the throughput of the network. However, the application warns the operator in the network operations centre of these circumstances by means of an alarm message (a panel lights up, a printer message is issued, or an indicator is activated on the operating console) and can request assistance from the operator to choose between various possible options for action.

Communications management layer
The communications management layer exchanges messages with the equipment via the standardised Q3 interface. This layer provides a standard interface for interrogating equipment. It is a gateway which translates standard requests from the system into the proprietary language format appropriate to the equipment type. It writes log files, including the results of interrogations or alarm messages raised by network components.

7.3.4 Maintenance

Maintenance procedures aim to:

- rectify network problems with the minimum of delay;
- evolve from network process measures that rectify problems to ones that prevent them from occurring in the future.

For this it is necessary to analyse all the available information in order to detect degradations before they cause an incident or fault.

7.3.5 Management of the Network

The objectives of evolution management are:

- to adapt products and services to meet customer needs;
- to adapt the network to the carried traffic, the network equipment, and the operational and maintenance practices.

Mobile Terminals

8

8.1 Overview of Mobile Terminals

A mobile station (see Figure 8.1) is made up of two physical parts; the terminal handset, which is the commodity hardware element, and the Subscriber Identity Module (SIM), which is personalised and unique to the subscriber. These two elements will be described in turn. Figure 8.1 shows a schematic of a portable terminal.

This chapter will describe the various categories of terminal and their mandatory and optional functional attributes. ETSI recommendations 02.06 and 02.07 define the types and characteristics of mobile stations.

A mobile terminal carries out two roles:

- a subscriber telephone;
- a means of connecting to a radio network.

Subscriber handset

The mobile terminal should have the same qualities as a domestic telephone, based on the following criteria:

- acoustic performance;
- the aesthetics of the handset;

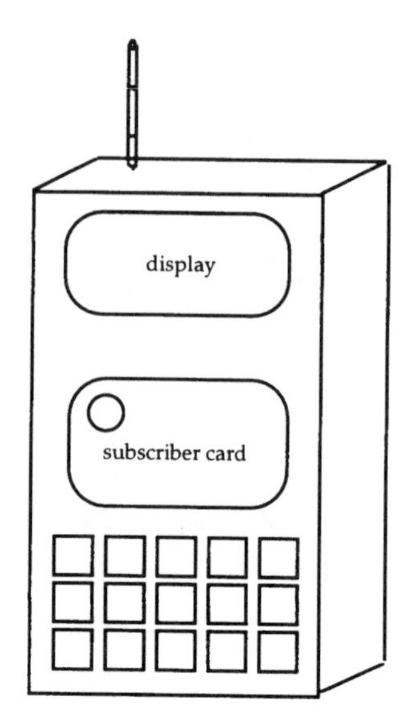

Figure 8.1: Portable terminal

- user-friendliness.

In addition, it is specified to be compact, lightweight and discreet, and it should consume little energy, in order to have as long a period of standalone operation as possible.

Radio connection
In its capacity as an interface to a radio network, the terminal should:

- conform to the electromagnetic compatibility standards;
- manage the signalling protocols;
- manage the encryption algorithms;
- provide digital speech processing.

To achieve low power consumption, and maximum stand-alone operating time, the terminal manufacturers have designed circuitry making use of low-power integrated circuits. However, the terminal design also takes account of the cycles of usage and idle periods, when the power consumption is reduced to a low level. For each situation, three states are defined:

- active-the circuitry is powered-up and operating;
- rest-power supply to the circuitry is completely cut off;

- standby-the clocks in the digital circuits are switched off, and the supply to the analogue circuits is disconnected.

The principal processor in the terminal controls changes of status of circuits which are activated in accordance with the necessary processing.

The GSM standard provides radio output power control on radio links. To achieve this, the base station continuously monitors the error rate on the data stream from a mobile terminal, and requests an adjustment of the radio power via signalling messages. This optimisation process contributes to the saving of energy. The transmit and receive cycles of a terminal (a time slot interval in a TDMA frame, or 577 μs every 4.615 ms) are also taken into consideration by energy economy algorithms in order to minimise consumption.

8.2 Types of Terminal

The standard defines three sorts of terminal:

- terminals fixed to vehicles;
- terminals that can be carried;
- hand-portable terminals.

Fixed terminals
Fixed terminals are normally mounted as permanent fixtures in the cabs of vehicles (lorries, cars, boats and trains) and on motorcycles, with the aerial sited externally, and using the vehicle battery as the power source. For road vehicles, the transceiver of the terminal is often mounted out of the way, in the boot or other storage space. For maximum driver safety, there are hands-free operation versions, which allow conversation to take place without letting go of the steering wheel (or handle bars). In the hands-free system, a small microphone can be mounted on the sun-visor, and a loudspeaker can be integrated into the handset or installed in the dashboard. Another option, which allows the handset to be used as simply as possible, is the facility to be able to take an incoming call by pressing any of the keyboard buttons, a simple locking code preventing unintended operation. Some models may be linked to vehicle-mounted fax machines, and top-of-the-range versions operate by voice-command. Fixed terminals that have their own battery may be converted to portable ones if they can be unclipped from their mounting.

Demountable terminals
These are similar to fixed vehicle-mounted types, but they can be trans-

ported outside of the vehicle. The telephone handset is linked to the terminal unit, which can be carried in a bag. These mobile terminals are designed for users who move around industrial sites, or in buildings, but who wish to remain in contact with the outside world. Hands-free versions of these terminals are also available.

This type of terminal is appropriate to different user situations, and they are usable within or outside vehicles. The power source can be either the vehicle battery or a battery integrated into the terminal, and the terminal aerial is not physically integrated with the vehicle. For a weight of less than 3 kg, the battery has a discharge time of 7 h in standby mode, and 2 h in continuous communication.

Hand-portable terminals

Hand-portable terminals are completely stand-alone, very compact (having a maximum volume of 900 cm^3), and very light (having a maximum weight of 800 g including the battery). This battery provides autonomous operation which varies according to the characteristics of the battery and the mode of operation of the terminal (stand by or communicating). The displays are smaller than in the other types of terminal (two lines of 16 characters). Hand-portable terminals can be mounted in vehicles, and have sockets to allow their batteries to be recharged, and to allow connection to external aerials.

The whole terminal is integrated into the body of the handset to save space and weight, but because these are lower power devices, they operate less well in buildings, although they are popular because of their wide mobility. Two varieties exist, one having a rigid non-retractable aerial, and the other having a flexible yet retractable aerial. The former type is larger, heavier and more robust, and more appropriate for frequent manipulation of the terminal, while the latter type is smaller, lighter and more fragile, and should be handled with care, but can be slipped into a handbag or a pocket.

8.3 Terminal Output Power

The main power characteristics of terminals are shown in Table 8.1. The mobile terminal output power can be stepped down in 2 dB increments by remote control from the base station.

Table 8.2 lists the principal features of GSM.

Table 8.1: Power classification of terminals

Class	Peak power output (W)	Terminal type
1	20	Vehicle-mounted and portable
2	8	Vehicle-mounted and portable
3	5	Hand portable
4	2	Hand portable
5	0.8	Hand portable

Table 8.2: Main GSM characteristics

General parameters	Value
Downstream frequencies	935–960 MHz
Upstream frequencies	890–915MHz
Radio channel spacing	200 kHz
Duplex spacing	45 MHz
Radio emission power	13–39 dBm in steps of 2 dBm
Data rise and fall time	28 μs
Parasitic emissions	< -36 dBm
Phase error	5° RMS
Transmission frequency error	95 Hz
Receiver sensitivity	104 dBm
Co-channel rejection	Signal -86 dBm
Intermodulation rejection	Signal -100 dBm
Signal blocking level	100 dBm

8.4 Mobile Terminal Characteristics

Two categories of terminal characteristics are defined:

- mandatory;
- optional.

Mandatory characteristics for a terminal type are required to be implemented in the instrument, while optional features implemented are chosen by the manufacturers. Table 8.3 lists the basic features of terminals.

The key for obligatory features in Table 8.3 is as follows:

0*: the feature is mandatory when there is an adequate man-machine interface;

Table 8.3: Basic terminal characteristics

Characteristic	Type
Direct international access	1
Called number display	0*
Automatic calling	0
Emergency calls	2
Self-test	0
Keyboard	1
On-off function	N
IMEI	0
Call processing indication	0*
Short message saturation indication	N
Service indicator	0
Indication of incorrect PIN code	0
Analogue interface	N
DTE/DCE interface	N
ISDN 'S' interface	N
Short messages	N
Name of PLMN network	0*
PLMN network selection	0

1: the means of entering characters 0–9, +, *, and # can be

- keyboard
- voice
- DTE interface
- any other means

2: emergency calls can be made with teleservice 12.

Automatic call-back of a dialled number can be invoked in three cases:

- the called number is busy;
- the called number is temporarily unavailable:

  no answer
  the number requested is out of service
  channel unavailable
  network resources unavailable

- the called number is unavailable for another reason

  the number dialled is not in service

the number dialled is incomplete
the network is out of service.

Automatic call-back for the first two reasons can be repeated up to ten times by operating a repeat button, but a minimum delay between repeat attempts of between 5 s and 3 min must be left. When the maximum number of repeat attempts has been reached, the repeat function shuts down, until a new number is dialled.

By displaying the dialled number, the user has some control over the dialling.

Call-progress indication takes the form either of tones or of a display of information provided by the network. For data transmissions, this information is sent to the data terminal.

The country code digits identify the Public Land Mobile Network (PLMN) identifier to which the terminal is connected. This information is necessary to the user when he needs to verify the name of the network when several networks are accessible from where he is situated, and he is making use of the 'roaming' option. The country and PLMN identification codes are therefore mandatory. The means of selecting a PLMN network are set out in the GSM standard 02.11.

Each terminal has an individual identifier International Mobile Station Identifier (IMSI), which the terminal communicates to the network when requested. The code-protected SIM module within the terminal contains the International Mobile Equipment Identifier (IMEI). When the SIM module is taken out, any call in progress is cut off, and any new call attempt is blocked, except for emergency calls.

An alerting signal indicates when there are messages waiting in the letter box located in the message management centre, and message originators can check the status of their messages (e.g. waiting or delivered). The mobile terminal user requests the transfer of messages when the alerting indicator is active, and acknowledges receipt to the message centre when they arrive. For short message users, another warning signal is activated when messages are refused because the memory is full.

Mobile terminals are equipped with a standard connector that provides an interface to a data terminal, a standard connector that provides an interface to an ISDN terminal, and an analogue interface connector to connect to a hands-free terminal.

The international access function gives standard access whatever network the terminal is currently located in, and for this the terminal is equipped with a button marked ' + ', which initiates a call to the international gateway. This function is available either directly, or as part of a memorised code in the terminal. This short-form code is useful because international access codes vary from one country to another.

An on-off switch allows the user to economise the battery usage of the terminal. The software switches the terminal status and before it enters the 'off' state, the terminal indicates this transition to the network. The PIN code may be requested before it can return to the 'on' state.

An indicator advises the user in a given zone that service is available and that he may use his terminal.

Table 8.4 lists some optional features of terminals.

Table 8.4: Optional features

External alarm
Call-charge counter
Earpiece loudness control
Second handset
User identification
Channel quality indicator
Hands-free operation
Selectable number of re-dial attempts
Short-code dialling
Outgoing call restriction
Multi-user terminal
Time-out before shut down

Short-code dialling looks up a subscriber telephone number stored in the memory and sets up a call to this number.

Outgoing call restriction has several available options:

- barring of outgoing calls;
- only stored numbers are authorised;
- only stored prefixes are authorised;
- local calls;

- national calls.

Switching to the out-of-service state can be delayed by a timer. For example, after switching off the vehicle engine the terminal will turn itself off automatically after a programmed delay.

An external alarm is a device that is separate from the terminal and that can signal the arrival of a call by means of a bell or light.

A call-charge indicator provides charging advice received from the network for the call in progress. The user can enter a command to ascertain the cumulative charge for each of the networks used, and for the previous call.

Supplementary services, for which there are defined procedures, are controlled by the user from his terminal.

The mobile station can serve several users for both incoming and outgoing calls, and it has a card reader for each user.

8.5 Functional Architecture of the Mobile Terminal

The terminal is comprised of five sub-modules:

- the radio sub-module;
- the processor sub-module;
- the synthesiser sub-module;
- the control sub-module;
- the user-interface sub-module.

Figure 8.2 shows the functional architecture of a GSM terminal

Radio sub-module
The radio sub-module controls transmission and reception. The receiver unit filters and amplifies the input signal from the aerial, and the transmitter unit generates, modulates and amplifies the broadcast signal.

Processor sub-module
The processor sub-module carries out the analogue-to-digital speech conversion, the signal demodulation and the channel coding and decoding.

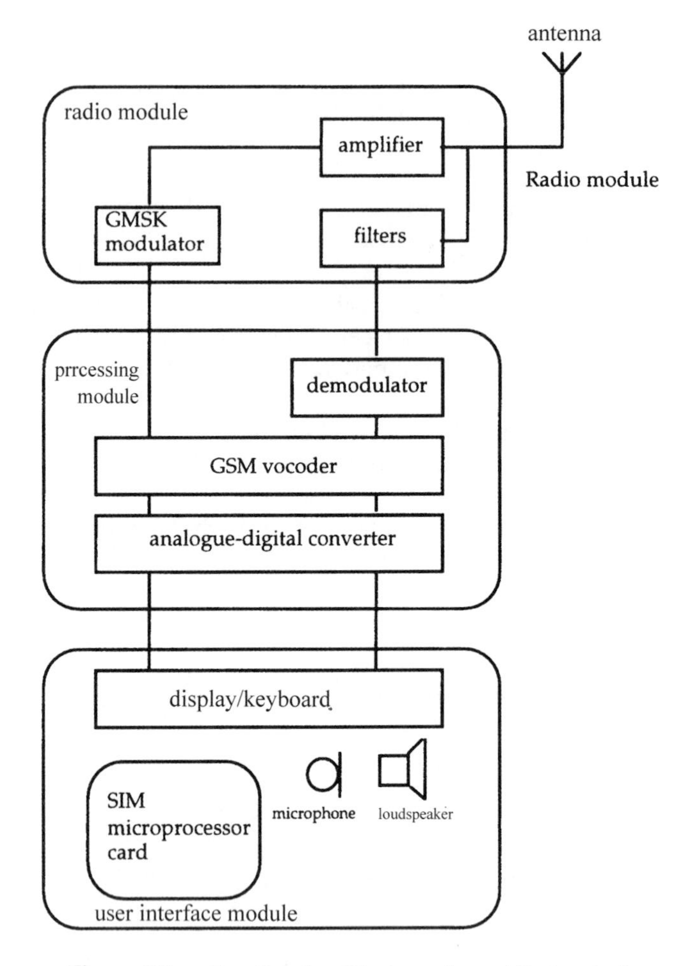

Figure 8.2. Functional architecture of a mobile terminal

Synthesiser sub-module

The synthesiser sub-module generates the transmitting and receiving frequencies and switches them at the TDMA frame rate.

Control sub-module

The control sub-module governs terminal operation and encodes the transmitted data. It manages the user interface (keyboard, display, SIM and buzzer) and the battery, under software control.

The user interface

The user interface comprises the following components:

- microphone, earpiece/loudspeaker;

- buzzer (bell);
- display;
- keyboard;
- SIM module.

8.6 Encryption

8.6.1 Encryption Concepts

Security methods are described in the CCITT X.400 and X.500 series of standards, and are largely based on public key ciphering techniques. The following sub-section briefly describes the principle of public key enciphering.

8.6.2 Public Key Ciphering

Public key encryption differs from usual forms of encryption in the dissimilarity of the means of enciphering and deciphering of messages. In classical encryption, knowledge of the encryption key F allows the decryption key F^{-1} to be calculated, which is not the case for public key encryption. In the latter system, a pair of keys, A and B, are generated by the originator, who conveys the public enciphering key A, and keeps the deciphering key B secret. The main characteristic of the public key enciphering method is that the deciphering key cannot be calculated from the public enciphering key. When a confidential message is transmitted, the public key provided by the originator is used to encipher the message, and the message can thus be kept confidential.

8.6.3 Authentication

Authentication gives the receiver of the enciphered message the means of identifying the origin of the message, i.e. the transmitter identity. Message authentication uses the principle of digital signatures, and the public key enciphering mechanism provides a means of generating a digital signature.

Consider the following hypothetical case:

- If a section of an encrypted message can be deciphered using a given key

Ks, then the message was enciphered using a dual public key Kp.

- If a particular person knows the public key Kp, then he must have been the author of the message.

- The message section is long enough to be distinguishable from a random sequence.

- The corollary of these hypotheses is that, given that only the sender knows the key Kp, he is the sender and the author of the message.

- Therefore a message received consists of a message and a section of the message enciphered using the senders secret key. To verify a signature, the receiver decrypts the enciphered part and compares the text obtained with the non-enciphered part in order to find out if the two agree and on condition that the message is of sufficient length.

8.6.4 Certification

Public key enciphering provides a means of maintaining confidentiality of communications, and also of authenticating the author of the message. However, it is equally important to be certain that a given public key belongs to a given user, and not to a usurper. A method is therefore needed for verifying the key and user as a pair. This is the mechanism of certification defined in the X.509 standard.

An 'off-line' system exists to generate a certificate for a user of the public key encryption mechanism. The aim of the certification system, as far as the user is concerned, is to generate a digital signature. The certificate is controlled by hashing its contents, and then comparing the result with the encrypted signature using the public key of the certification system, If the comparison is true, the user is certain that the certificate has been generated by an authorised system. Therefore the receiver of the message knows that the certificate has been established by a system that assures the validity of the public key-sender combination of the message.

8.7 The Subscriber Identity Module (SIM)

8.7.1 Overview

Two separate physical entities make up a mobile station:

- the mobile equipment (ME);
- the SIM.

Figure 8.3 shows an outline of the terminal.

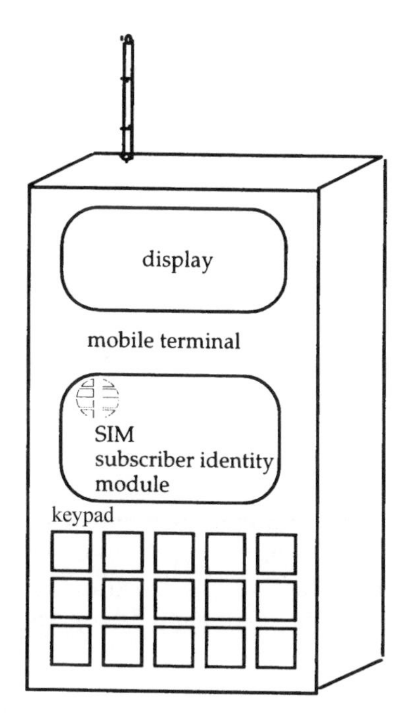

Figure 8.3. A schematic of a mobile terminal

The GSM recommendation 02.17 specifies both the interface between the mobile terminal and the SIM, and the logical structure of the SIM seen by the mobile terminal. The recommendation guarantees the compatibility of SIMs with mobile terminal equipment from different suppliers.

The mobile terminal structure

The mobile terminal equipment is the commodity item in the mobile station. The tangible element is the ensemble formed by the radio equipment, the keyboard, the display, the microphone and the earpiece/loudspeaker, which provide the subscriber with the physical means of accessing the network, while the hidden part is the associated software. The mobile equipment can be considered as the mobile terminal minus the SIM, and is the hardware which enables a user to set up connection to the network and to communicate with it.

The subscriber identity module

The SIM is the personalised part of the mobile station and operates in conjunction with a memory card, which separates the means of using the network from the definition of the access rights of the subscriber. The SIM fulfils the security needs of the network operator and the subscriber in so far as the operator controls the validity of a subscription at all times, and the subscriber wishes to protect the privacy of his personal life and activities.

Two simultaneous roles are embodied in the concept of the SIM: on the one hand it is a logical functional entity, and on the other it is a physical storage device holding the subscriber's data. In addition, the data contained on the card is protected by logical keys. The microprocessor card holds several categories of information, and each category has a particular code protecting the information from being accessed illegally. The user must always identify himself, and this requires several code keys. The portability of the information in the card memory provides significant flexibility of use for the subscriber, the network operator and the terminal manufacturer, and opens up the possibility of standard mass produced terminals. The particular SIM – mobile terminal combination used is not dictated by network design. A subscriber can use different terminals as he wishes, according to his needs. As far as the network operator is concerned, the SIM embodies the subscription account. It completely identifies the account owner and can be modified, or the subscription options changed. Network operators manage SIMs independently of the terminals.

8.7.2 Subscriber Identity Module Functions

The most important of the many SIM functions are user authentication, radio transmission security and storage of subscriber data.

Subscriber authentication

The mobile station can access the network if the network has previously authenticated the user. The authentication function is shared between the network authentication centre and the SIM card.

Two techniques of authentication of one entity, A, by another, B, exist; one active and the other passive.

The passive authentication technique: code word

The basis of this technique is the sharing of particular knowledge between entities A and B. The knowledge takes the form of a secret code word, known in principle only to the two entities.

The scenario is that entity A chooses a secret personal code, which is then communicated to B. When A wishes to be recognised by B on a particular occasion, it supplies B with the code word. Entity B compares the code word received with what it already holds, and if it corresponds, then B recognises A. This is a static technique, and the code word is invariable. B recognises as A any entity able to supply the correct code word. This is illustrated in Figure 8.4.

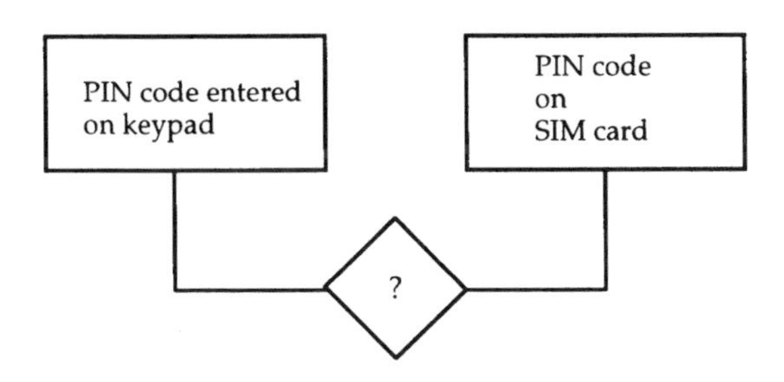

Figure 8.4: Code verification

GSM recommendation 02.17 specifies the use of this technique for the purpose of consulting the information stored on the SIM card; for example, the list of messages received. The Personal Identification Number (PIN) code has four digits, which can be changed at the discretion of the subscriber. The facility of checking this personal code can be disabled by the subscriber at any time, but the network operator has the choice of whether to provide this option. After three unsuccessful authentication attempts have been made, using the wrong code, the SIM is disabled automatically and the terminal cannot be used to connect to the network with the SIM in this state. In order to restore the card to a valid state, a release key code (composed of eight digits) is required. It should be noted that the mobile terminal does not have any power to access the network solely on basis of the data held in the equipment. The data contained in the network databases are references to enable access rights to network services.

The active authentication technique: proof
In this technique, the principle is the provision to B by A of a sign of acknowledgement which B alone recognises as 'proof', which A will recognise automatically without knowing it. To instigate an acknowledgement, A proves that it is in possession of the information that B has communicated to it by giving a correct reply to a particular request. In this technique, B makes a different request each time that an entity presents itself for recognition. The request–response combination is therefore dynamic. The proof is an algorithm held in a secure store.

With the code word technique described above, an unauthorised outsider could eavesdrop on the dialogue between A and B during the recognition phase, and then masquerade as A when signalling to B, who would be unable to tell that impersonation was being attempted. This type of attack is not possible in the proof technique, where the criteria for recognition vary according to the request made, and knowledge of a single response has little chance of being effective.

GSM uses the proof technique for each request for service made by the mobile station. The SIM card contains the 'proof' to be supplied to the network on each occasion that it is demanded. Figure 8.5 shows schematically the authentication mechanism of a subscriber at the time of his entry on to the network. The secret key, Kp, and the enciphering algorithms are stored on the SIM card and in the HLR, and the network operator allocates and writes the key Kp on to the card. This key is a particular enciphering algorithm, and it constitutes the proof for the network operator. However, the card holder never has access to this information.

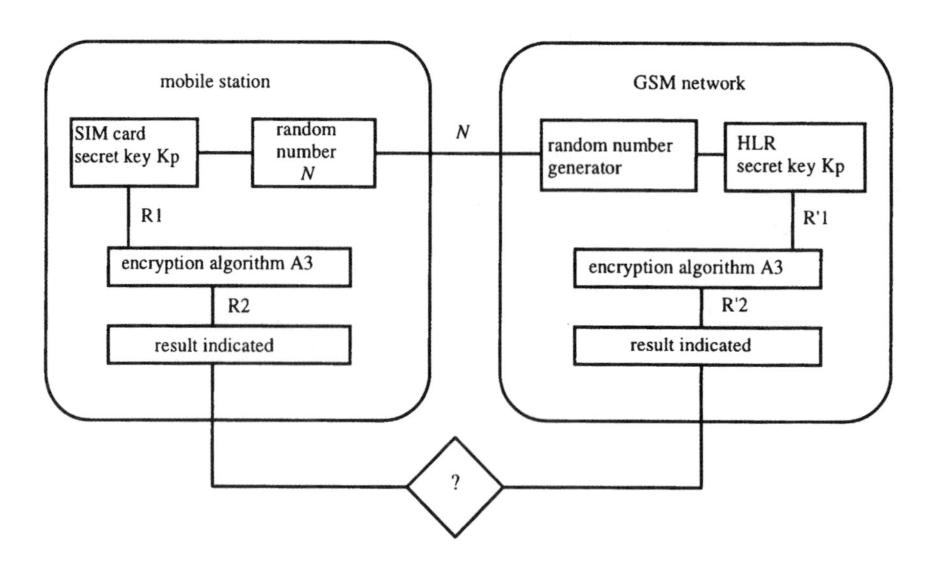

Figure 8.5. Subscriber authentication mechanism

When a subscriber requests network service, the network initiates the subscriber authentication process, as follows. In the Authentication Centre (AUC), a random number generator sends a number N to the mobile station. In the mobile station, the number is encrypted using the key Kp. The result of this encryption, R_1, is also encrypted, this time using a second algorithm key, A3. The result of the latter encryption, R_2, is sent to AUC and compared with the expected result. If there is correspondence, the subscriber is permitted to use the service requested.

Although the subscriber authentication process is standardised, each network operator is free to choose its own encryption algorithms. The SIM card holds encryption algorithms used for transmission on the network (Ax) as well as the personal key Kp (which is another encryption key). However, the results of calculations are never communicated in the clear around the network, for reasons of security.

The authentication mechanism implemented by the system is shared between the SIM and the AUC, and requires the retrieval of data from the HLR database. For obvious security reasons, access to the building accommodating the HLR is restricted, and the databases are centralised for the same reason.

Security

The radio channel is the most vulnerable part of the network, and security features are implemented to protect the subscriber and the network operator against illegal use by an outsider, and by eavesdroppers on the channel. The security features address the confidentiality of the subscriber identity and authentication process, and the confidentiality of the signalling and personal data conveyed over the network channels.

Three types of security measure are implemented:

- the SIM is authenticated by the system, which protects the network operator against use by non-registered users;
- the subscriber identity is protected and never conveyed openly on the network;
- the radio link between the terminal and the network is encrypted to avoid eavesdropping.

Authentication of a subscriber

A permanent personal IMSI number uniquely identifies subscribers in all GSM networks, and this number is universal in all PLMNs. Subscribers also have listed directory numbers which are not the same as the IMSI, and which are the numbers that are dialled in calls.

The IMSI number embodies the following subscriber information:

- country;
- home network in the country;

- Home Location Register (HLR);
- location of the subscriber details in the HLR.

Other numbers are attributed to the subscriber, either permanently; such as the directory number, or temporarily, such as the Temporary Mobile Subscriber Identity (TMSI). This data, and also the necessary correspondences, are maintained in the HLR and VLR. The enciphering algorithms are also used in the network operations terminals in the HLR and VLR to keep operations and signalling information peculiar to subscribers secure. ETSI has defined a set of algorithms for this purpose, but the network operator is free to choose those employed in his own network.

Measures taken to protect subscriber identity

In order to keep the identity of the subscriber confidential, the transmission of his IMSI openly on the radio channel is strictly limited, and the TMSI number is substituted where possible. A TMSI number is ascribed to a mobile terminal within the limits of a geographical zone, and for a sufficiently short period that any intruder would have little chance of finding and exploiting the correlation between the TMSI number and the IMSI. The TMSI number is held on the SIM card and in the VLR.

Protection of data on the radio channel

To guarantee the confidentiality of information transported on the radio channel between the base station and the mobile station, the signalling and speech data are encrypted using an encryption algorithm. After authentication of the subscriber, the network supplies the mobile station with a first enciphering key, Kc, when it connects to the network. The enciphering key used with a particular frame is calculated in the mobile station using a formula associated with the authentication algorithm, based on a random number and the authentication key. The enciphering algorithms are common to all mobile terminals and all base stations. Figure 8.6 presents the mechanism for encryption of frames used by the mobile station and by the base station.

Three categories of data are stored on the SIM:

- permanent identification data;
- temporary identification data;
- additional data.

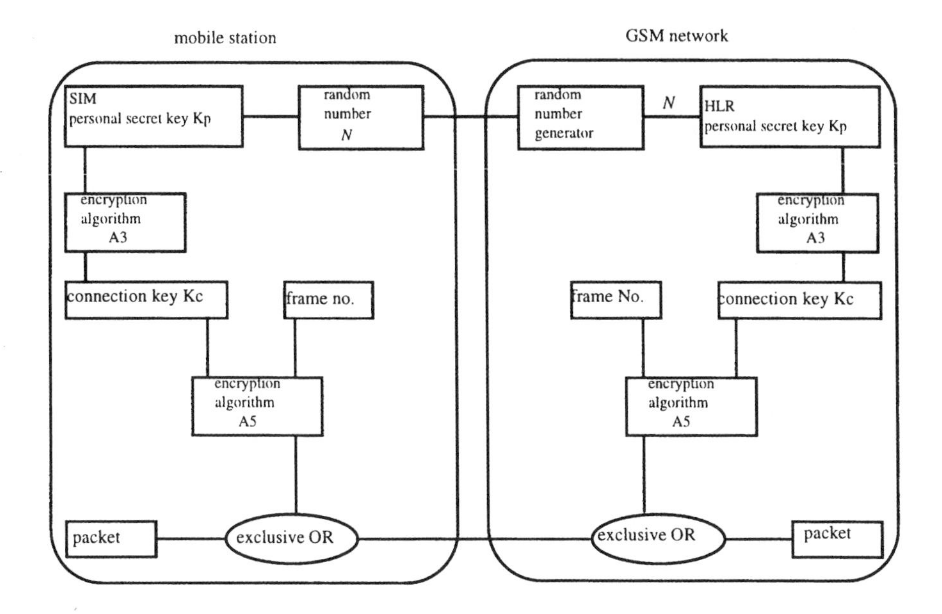

Figure 8.6. The frame encryption mechanism

The permanent identification data are:

- the IMSI;
- the authentication key (Kp);
- authentication algorithms;
- calculation algorithms.

The temporary identification data are:

- the TMSI;
- the terminal location.

The additional data are:

- list of short-code directory numbers;
- short messages;
- list of fixed programmed numbers in the case of usage restriction (outgoing calls, international calls);
- tariff information supplied by the network.

Short code dialling management
The SIM card contains an editor to allow access to, and listing of, numbers stored in the short-code dialling memory. In order to create an abbreviated number, the subscriber inputs a mnemonic code, and associates it with a full directory number. If other applications are to co-exist within the SIM using the same card medium, it would be simpler for the user to have a menu that is common to these applications.

Short message storage
The short message delivery teleservice enables mobile stations to receive short messages broadcast by the network. It is often convenient to store these messages in memory for later reading, and some mobile terminals offer this facility.

Call restriction to fixed numbers
Certain subscription types only allow calls to be made to fixed pre-determined numbers (for example, to emergency service numbers). This function is thus a form of call filtering.

Charging advice
This facility informs users about the cost of the call in progress, so that they can monitor their telephone usage costs.

8.7.3 Technical Features of the Module

Implementation
The SIM card is an application of the microprocessor card defined by ISO under standards IS 7816-1 and IS 7816-2. Two implementations are defined, both of which offer portability:

- SIM card;
- Plug-in SIM.

SIM card
This is a credit card sized memory card dedicated to GSM. If the card is multi-service, the SIM is the particular application appropriate to GSM: if not, the module and the card may have overlapping functions. The user inserts the SIM card into the card reader equipment on the terminal in order to activate it for service.

Plug-in SIM
The GSM recommendation 11.11 specifies a mechanical interface between the SIM and the mobile terminal equipment. The remaining specifications are similar for both the card and plug-in versions. This version is intended

for hand-held stations, where high manoeuvrability is important. The design uses the same type of component as the credit card sized SIM, but is smaller.

In cases in which the equipment possess both types of interface, the card-version SIM has higher priority.

Physical description of the module

The ISO standards IS 7816-1 and IS 7816-2 specify the physical and mechanical characteristics of the microprocessor card supporting the SIM. Figures 8.7 and 8.8 show a microprocessor card and its electrical interface, a metal pad equipped with eight electrical contacts.

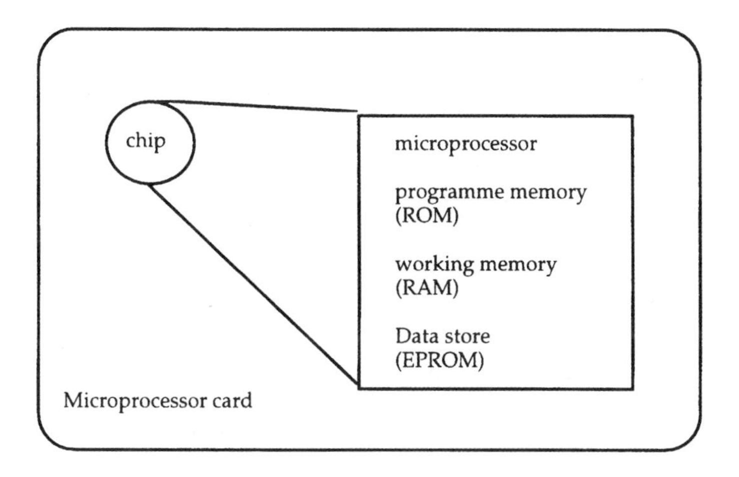

Figure 8.7: A microprocessor card

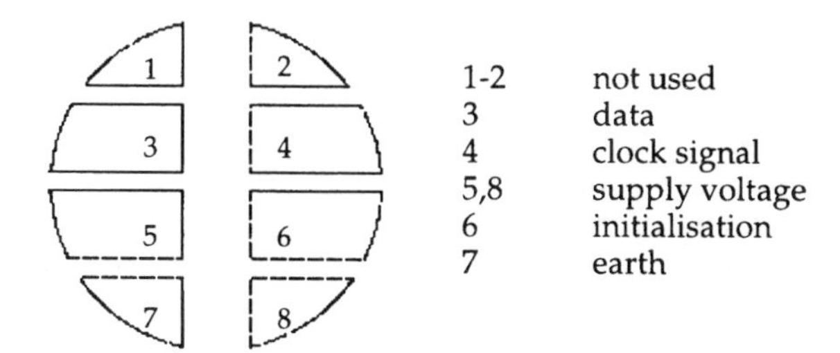

Figure 8.8: The electrical interface to the chip: contact layout

A metal pad fixed to the back of the integrated circuit chip protects it from thermal, electrical, chemical and radiation effects due to the external environment. The pad provides a standardised eight-contact external electrical connection interface from the chip. Three contacts (five, seven and eight) carry supply current, a further contact (four) connects to a clock source from an external equipment, which synchronises the card with this source, another contact (six) serves to set up signalling to the external equipment interfacing to the card, a further contact (three) acts as a data line, and the remaining two contacts (one and two) are reserved. Figure 8.9 shows the chip architecture schematically.

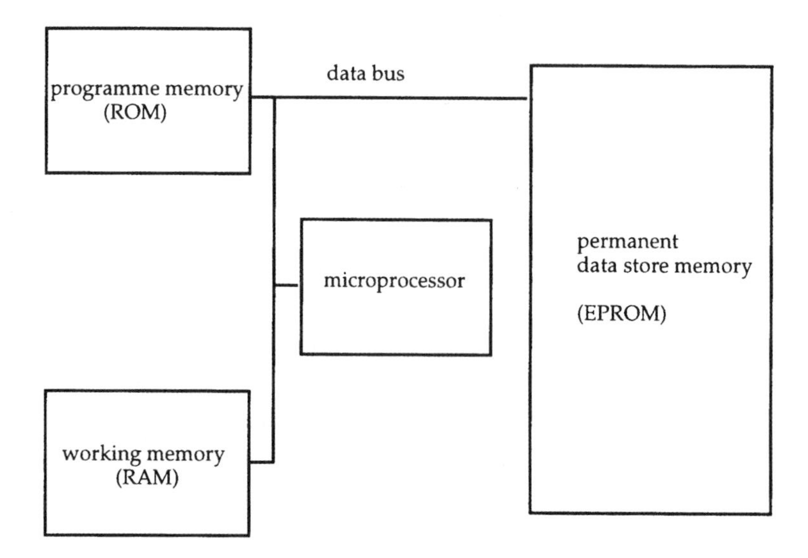

Figure 8.9: The functional architecture of the chip

The microprocessor chip consists of the following elements:

- A monolithic self-programmable microprocessor (μP) is the data processing unit of the chip.

- The programme memory. The Read Only Memory (ROM) is mask-programmed at the fabrication stage. It contains the operating system of the card, which controls the microprocessor functions when it receives commands from the mobile terminal.

- The dynamic memory. The Random Access Memory (RAM) is the volatile working memory of the microprocessor.

- The EPROM memory. The Erasable-Programmable Read Only Memory (EPROM) stores both permanent data, and also electrically-configurable data.

- The data bus. The bus links the microprocessor to the various memory areas.

Signalling protocol between the SIM card and the mobile equipment

The signalling protocol has been developed entirely by the Subscriber Identity Module Expert Group (SIMEG) in ETSI. GSM recommendation 11.11, which is very similar to the ISO IS 7816-3 standard, defines this protocol. The protocol governs SIM-to-terminal equipment signalling and is an oriented-character asynchronous protocol, designated by the notation '$T = 0$'.

Organisation of the microprocessor memory space

GSM recommendation 11.11 describes the logical structure of the SIM memory. The currently available components have a limited memory capacity, but this is expected to increase in the future. The programmes have 5 kbytes of ROM allocated, and the dynamic data have 2 kbyte of EPROM allocated.

The EPROM memory stores all of the data acquired during the life of the card. It is divided into six distinct zones:

- The manufacturer zone. The chip is programmed with the manufacturer and the component number when the chip is manufactured. This zone is programmed when the card is personalised and includes the code defining the application, and the pointer to the other zones and the conditions of access to these zones.

- The secret zone: This zone contains data which cannot be read or overwritten. This data controls the following: access to the function activating the card, the sender key, the key (or keys) providing service (intermediate between the sender and final user) and the code-carrying key (confidential codes).

- Status zone: This zone stores the keys needed to access the protected data.

- Confidential zone: This zone holds the permanent protected data, which is only accessible via the sender key or the code-carrying key.

- The transaction zone. After the personalisation of the card, this stores data that can he accessed with or without a key, according to the conditions written into the manufacturer zone.

- The open zone. This zone contains the non-confidential data which may be freely read.

Each memory zone has a level of security allotted to it for reading or writing. As in any data retrieval system, some data is public and some is private, and so access to it needs to be controlled.

Recommendation 11.11 defines the concepts of data 'directories' and data files for identifying data written to the transaction zone, and data files are given a degree of security. The different types of secure operations are:

- reading (condition);
- writing (condition);
- invalidation (condition);
- restoring (condition);
- initialisation writing (condition);
- extension (condition).

The different sorts of condition to be fulfilled to enable an operation to be carried out include:

- never – the operation is barred;
- always – the operation is authorised for all users;
- PIN – the user must first provide his personal number;
- administration management (ADM) – the user must first provide the correct ADM code.

The highest-level directory is the root directory. Three directories have been defined:

- root;
- GSM (which contains the files specific to the application);
- telecoms (which contains the files relating to other facilities).

A file is composed of two parts, a header and the data content. The header contains a file type descriptor:

- name;
- length;

- security level;
- the type (binary or formatted).

There are two file types. These are:

- binary – the content of the file is a set of bytes, which are read directly;
- formatted – the content of the file is structured into fixed length recorded sectors. At the creation of the file, the number of recorded sectors is fixed, and transactions with the file are carried out by writing to the sectors. The sectors are arranged sequentially, each sector being accessed through an index.

File transactions include:

- creation – allocation of memory space;
- selection – on inputting the name, access to the header descriptor;
- extension – to increase the size of the file;
- invalidation – to prevent all actions on the file, except restoring it;
- initialisation writing – first write operation on a created file;
- updating – erasing followed by writing;
- reading – access to data stored in a file;
- sorting – searching a file.

These actions are subject to the restrictions imposed by the security level allocated to a file when it is created.

Authentications

The main function of the SIM is to provide authentication of itself to the network. A mobile station cannot access the network service unless it has all of the data relating to a subscriber, and therefore its SIM. However, other sorts of authentication are necessary in order to execute restricted actions; for example, accessing a protected data file requires the submission of a secret code.

8.7.4 Module Life-cycle

The life-cycle of a SIM is the period of time separating the manufacture of the chip, which supports the SIM, and the withdrawal of the SIM from service. The GSM specifications identify two periods:

- GSM
- ADM

GSM

The GSM period is the period of time during which a SIM is allocated to a particular subscriber, and the SIM has a valid IMSI. During this period, all operations between the module and the terminal equipment conform to the recommendations 02.17 and 11.11 of the GSM standard.

The following network operations occur during the initial phases of a call:

- connection to the network;
- call set-up;
- conversation;
- clear down;
- change of location.

The SIM provides the following services when it is installed in a mobile terminal equipment:

- storing the security information in memory (IMSI, keys, TMSI, LAI, Kc, etc.), authentication of the subscriber and enciphering of data;
- management of the PIN and verification of the PIN;
- management and storing of subscriber data (service set, messages, charge advice, etc.).

Temporary subscriber information is stored on the SIM card at the end of each call, and in the mobile terminal.

ADM

Administration periods fall into the following stages:

- manufacture;
- distribution;
- pre-personalisation;

- personalisation (recording of the subscriber identity (IMSI), key authentication, etc.);
- invalidation.

These operations take place in the context of relations between the customer and the supplier, but also directly concern the network operator.

They are not the subject of standards. However, the GSM recommendations 02.17 and 11.11 give pointers aimed at harmonising these operations. The only actions dealt with by the standards are those concerning the management of subscribers in several networks.

When the SIM module is contained within a multi-service card, the private GSM data and programs are kept together in a file named GSM.ADF.

The other services present in the card do not have access to private GSM information. For this a mechanism present in the card verifies the access rights to the GSM.ADF file.

The GSM.ADF file contains headings from which the management recognises the following entities:

- manufacture – serial number, encryption algorithm;
- SIM distribution – creation of the GSM.ADF file;
- retailer – control of the key Ki, the IMSI, the directory number, network access authorisation and the initial PIN number;
- service provider – management of user data;
- the subscriber.

Management of the PIN number is carried out as follows. The PIN is a number containing from four to eight digits. The subscription retailer installs the first PIN number, but the user can change it as often as he wants afterwards. Using a special function, the user decides whether or not he wishes to use the PIN, and he can change his mind later.

A warning signal on the terminal advises the user that an incorrect PIN has been entered, and after three incorrect attempts the SIM is disabled.

The administration management (ADM) stages of SIM life-cycle

The following six types of operation take place during the ADM part of a SIM life cycle:

- Manufacture. The manufacturer delivers cards containing all the necessary functions, the algorithm, and the authentication mechanism. It then remains to initialise the cards to render them operational, and for that an authentication key and an IMSI number have to be recorded on each card.

- Distribution.

- Pre-personalisation. Pre-personalisation is the action of writing an authentication key and an IMSI number into the card memory.

- Personalisation. Personalisation is the action of writing a directory number into the memory of a pre-personalised card, which occurs when an account is opened. This action requires a personalisation key from the network operator.

- Disable/enabling. A card is disabled after three unsuccessful attempts at entering a PIN number. A disabled card cannot access the network, and in order to re-enable the card, a special eight-character key is required.

- Re-personalisat ion. A card may be re-personalised by a network operator. A dedicated re-personalisation key is required for this, and the procedures then follow the same sequence as for the pre-personalisation and personalisation.

8.7.5 Data Stored on the SIM

Subscriber information

Subscriber data is written on the card at a well defined point in the card's life-cycle. This is listed below:

- the card serial number (this identifies the card, the manufacturer and the operating system version, cf. ISO 7812);

- the card status (active or disabled);

- the service code (e.g. GSM);

- the pre-personalisation and personalisation data (directory number);

- the re-personalisation data;

- the authentication algorithm parameters;

- the authentication key;
- the IMSI;
- the pre-personalisation and personalisation keys;
- the enciphering key;
- a sequence of numbers for encryption;
- the TMS;
- the LAI;
- subscriber information;
- temporary location information;
- the validated status (updated);
- the list of barred networks (a maximum of four networks);
- network access class;
- the status of the PIN activation/de-activation function (authorised/barred);
- the PIN status (in service or out of service);
- the PIN;
- the PIN error counter;
- the re-enabling key;
- the status of the inter-PLMN roaming function (authorised or barred).

The following data is loaded when pre-personalisation occurs:

- the re-personalisation data;
- the card status (barred);
- the IMSI;
- parameters relating to the authentication function;

- the authentication key;
- the PIN activation/de-activation function (authorised or barred);
- the initial PIN number;
- the re-enabling key;
- the initial value of the PIN error counter (which equals 0);
- re-personalisation key.

The personalisation operation writes the following data to the card:

- the subscriber data;
- the personalisation data (date subscription account started, limit of validity);
- the network access class

0–9	for ordinary subscribers
11	network operator
12	security services
13	public utilities (water, gas, etc.)
14	emergency services
15	network operator personnel.

The mobile terminal equipment manages the following information:

- the enciphering key;
- the TMSI;
- the LAI;
- the GSM service information;
- the temporary location information;
- updating of the status of the module (in service, disabled);
- the list of barred networks (a maximum of four networks);

- the network access class;
- the sequence number for enciphering.

General data
The SIM module stores permanently the following data relating to the GSM services or to the supplementary services:

- short messages received;
- charging data;
- a directory;
- a list of authorised directory numbers (if call restriction is set up);
- the status of out-going calls (authorised or barred);
- the PLMN list.

Other data may be stored if required.

8.7.6 'MIMOSA' a Multi-Service Card

The French Centre for combined study of telecommunications and broadcast (CCETT) has designed a multi-service card called MIMOSA (an acronym arising from the French for 'Secure multi-operator microcalculator with protection by accreditation'). It uses a public key encryption technique based on a 'zero knowledge' algorithm. This card simplifies the method of making cards available and allows new services to be opened on the card without involving the card distribution service. The card has the means of conducting a secure transaction with a service provider, and allows the latter to convey the service key. The card distribution service is limited to the writing of a notice of accreditation on the card, and this could be carried out by the card manufacturer. The card-could be simply sold in a shop, since it initially has no rights enabled, and the owner can choose the service providers that he wishes to subscribe to.

The file organisation and the system architecture conform to the Iso 7816-4 standard. MIMOSA cards are able to authorise each other and to send and control signatures.

The Philips component 83C852SC is the first secure chip designed for public key encryption algorithms which can be integrated on to a card, and it is equipped with a calculation cell for processing specialised RSA type algorithms ('RSA' is derived from the names of the inventors, Rivest, Shamir and Adelman). The component also has an integrated 8-bit microprocessor 80C51, a 6 kbyte ROM, a 256 byte RAM and a 2 kbyte EEPROM memory.

The DECT 1800 Standard

9

9.1 DECT Technology

The Digital European Cordless Telecommunications (DECT) standard has grown out of the need to provide cordless communications, primarily for voice traffic, and also to support a range of other public traffic requirements. The standard has been designed to achieve this versatile support of applications at a cost that encourages wide take up. It is envisaged that DECT will provide personal telecommunication services in residential, neighbourhood and business environments. It is particularly targeted at the following applications:

- residential-domestic cordless telephones;
- public access services;
- cordless business telephones (PABX);
- cordless data-Local Area Networks (LANs);
- evolutionary applications(extensions to cellular radio, and extensions of the local public network).

One primary objective of the common interface standard is to provide for inter-operability between different commercial variants of equipment and to offer users a family of telecommunications services for voice and data, either as basic services, or with optional (and compatible) extensions. The standard provides escape routes that allow manufacturers to retain options for innovation and product differentiation. In addition, reserved codes have been included in the standard to build-in mechanisms for evolutionary development of the standard.

A connection is established by transmitting bursts of data in the defined time slots. These may be used to provide simplex or duplex communication. Duplex operation uses a pair of evenly spaced slots – one for transmitting, and one for receiving.

The simplest duplex service uses a pair of time slots to provide a 32 kbit/s digital information channel capable of carrying coded speech or other low-rate digital data. Higher data rates are achieved by using more time slots in the TDMA structure, and lower data rates can be served by using half-slot data bursts.

DECT is able to support a number of alternative system configurations, ranging from single cell equipment (e.g. domestic fixed systems) to large multiple cell installations (e.g. business cordless PABXs).

The DECT standard provides Dynamic Channel Assignment (DCA), which offers adaptive channel re-use, which removes the need for the cell frequency planning usually associated with cellular radio systems. In this process, the search for free channels to carry a new call is achieved by both the terminal handset and the base station in DECT DCA. The criteria for selecting correct channels are based on radio channel availability measurements at both ends of the link. Such an approach is essential for high density environments where the interference limitation is very important, and it also allows simpler system deployment planning. The use of DCA results in an interaction between different cordless links such that the performance of a cordless system must be derived from analysing the system as a whole rather than considering a link in isolation from its dynamically changing radio environment.

DECT uses 32 kbit/s ADPCM G.721 speech coding, low transmitter power, and time division duplexing, transmitting and receiving on the same channel frequency. DECT employs a 24 time slot scheme, in which a DECT carrier can support multiple calls through a single RF receiver.

An important feature of the DECT frame structure is the ability to use multiple time slots for one call in order to support higher data rate links, and so to increase the accessible bandwidth. If necessary, the occupied time slots can change carrier frequency within a single frame, so providing more flexibility to the system.

9.2 System Features

The standard seeks to provide an interface with ISDN terminals. However, the characteristics of radio communications are different from those of the fixed ISDN network.

Resource management
In radio communications, transmission channels are shared between users, and occupation of channels has to be managed as a function of time. This constraint does not exist in a fixed network, where the channel is allocated for an indefinite duration.

Mobility management
Cordless telephones are intrinsically mobile, and as a consequence there is specific functionality in the lower communication layers in order to manage mobility by tracking the movements of the mobile terminal. In addition, public networks need a network user authentication process to avoid arguments over billing for services.

Control of transmission errors
Radio communications are prone to errors and it is necessary to protect information by the use of coding and by use of appropriate protocols.

9.3 Technical Characteristics of the Standard

9.3.1 Frequency Spectrum Use

Two methods are employed to optimise the use of the frequency spectrum:

- segmentation of the frequency spectrum;
- dynamic channel allocation.

Frequency spectrum segmentation
The standard defines ten carrier frequencies. A geographical area is subdivided into zones, according to the general principles of cellular radio telephony. The zones are divided into cells, and a base station manages the communications within each cell: the smaller the cell, the greater the scope for re-using frequencies. The small distance between cells and the high traffic densities involved sometimes necessitate communication on two channels simultaneously when handing over between cells. The handover algorithms must be efficient and reliable, and this is accomplished by the mobile terminals themselves, which imposes only a light load on the network infrastructure.

Dynamic channel allocation

A dynamic allocation algorithm manages the allocation of traffic channels to network users. During a radio transmission, a mobile terminal only uses the traffic channel for some of the time: for the rest of the time it is exploring other channels in its cell and in the adjacent cells. The terminal stores this information, and uses it to control the handover if the user moves into another cell. It is the mobile terminal that both makes the decision to implement a transfer and chooses the new traffic channel, in conjunction with the base station, without breaking the communications link. At the end of this process, it releases the original channel when the link quality is better on the new channel. The dynamic allocation method means that the terminal is able to adapt to changes in the network as a result of new cells being added in order to increase coverage or to increase capacity.

Table 9.1 lists the key characteristics of the DECT standard.

Table 9.1: Characteristics of the DECT standard

Number of carriers	10
Carrier spacing	1.728 MHz
Peak transmitter power	250 mW
Frame length	10 ms
TDMA multiplexing	24 time slots
Number of duplex radio channels	12
Overall radio bit rate	1152 kbit/s
Overall bit rate of a speech channel	96 kbit/s
Synchronisation field	3.2 kbit/s per time slot
Field A (signalling, commands)	6.4 kbit/s per time slot
Field B (traffic)	32 kbit/s per time slot
Number of terminals supported	25
Telephony traffic capacity	5 Erlang
Compression algorithm	ADPCM G.721

9.3.2 Physical Channels

Channels are generated by duplex time division multiplexing, and Figure 9.1 shows the frame structure (cf. ETSI standard TER 015)

Each frame is 10 ms long and is divided into 24 time slots. The first 12 of these are reserved for communication from the base station to the mobile terminal; and the following 12 are allocated to communication in the other direction, giving a 12-channel duplex capacity. Each time slot (416.7 μs) contains a physical channel (364.6 μs), which conveys an information packet of 420 bits (388 payload bits and 32 synchronisation bits).

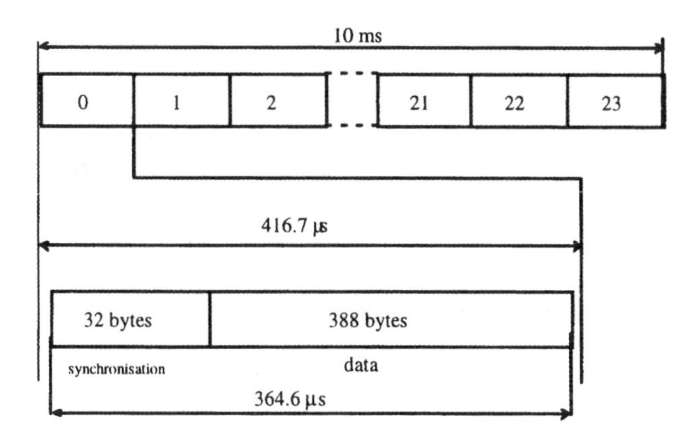

Figure 9.1: The structure of a DECT frame

The time difference between a time slot (416.7 µs) and a physical channel (364.6 µs), or 52.1 µs, accommodates the offsets between the clocks in the mobile and base stations and also the propagation time of the radio wave between the transmitter and the receiver. The overall bit rate of a physical channel is 42 kbit/s, but the usable bit rate is only 38.8 kbit/s.

Figure 9.2 shows the packet formats for the physical layer and the Media ACcess (MAC) layer, which is divided into four fields of 48, 16, 320 and 4 bits, representing the signalling, Cyclic Redundancy Check (CRC), useful information and X fields, respectively (cf. ETSI standard TER 015).

The ***signalling field*** (C/P/Q) has a length of 48 bits and transports signalling messages.

The ***CRC field*** carries the error control checksum.

The ***information field*** transports the application data, and gives a usable bit rate of 32 kbit/s.

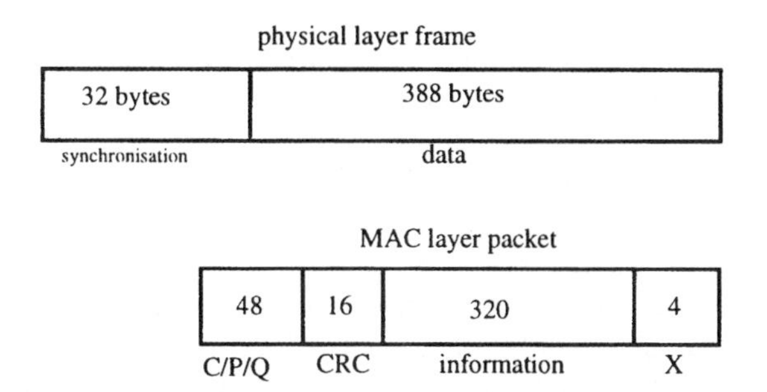

Figure 9.2: The DECT packet format for physical layer and MAC layers

The *X field* is used to detect and protect the information field from interference between two adjoining channels, and also in system synchronisation. The 4 bits in this field contain the control checksum of the 320 information bits. The receiving application calculates the checksum, and if it differs from the received *X* value it knows that there has been interference.

9.4 DECT Software Architecture

The DECT software stack has four layers, as shown in Figure 9.3.

The DECT stack is similar to the OSI model, with a MAC layer added as in local area networks.

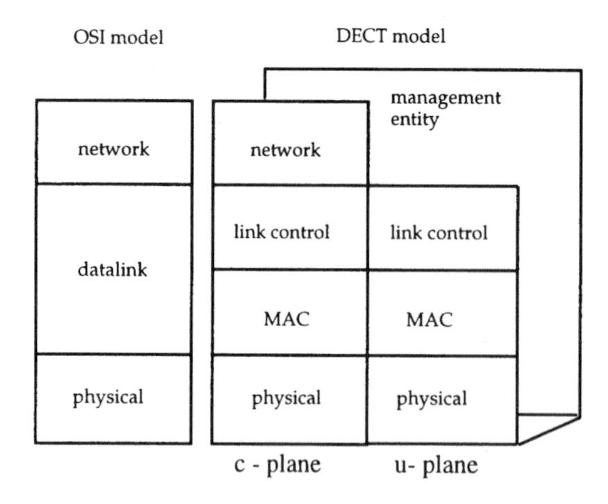

Figure 9.3: OSI and DECT Software Stacks

9.4.1 The Physical Layer

The physical layer divides the radio spectrum into channels. This division occurs in two fixed dimensions-frequency and time.

Frequency and time division uses Time Division Multiple Access (TDMA) operation on multiple RF carriers. Ten carriers are provided in the frequency band 1880–1900 MHz and on each carrier, the TDMA structure defines 24 time slots in a 10 ms frame, where each time slot may be used to transmit one self-contained packet of data. Each transmitted packet contains a synchronisation field together with control information, service information and error control.

The radio spectrum is also divided spatially into cells, in which the same physical channels may be re-used in different locations. Spatial re-use operates according to the principles of Dynamic Channel Selection (DCS).

Each fixed radio end-point operates according to a local timing reference and the physical layer is then responsible for transmitting packets of data under direct control of the MAC layer. Adjacent fixed points may be synchronised. This gives some advantages, particularly in high traffic situations.

9.4.2 The Media Access Layer

The MAC layer performs two main functions. Firstly, it selects physical channels, and then establishes and releases connections on those channels. Secondly, it multiplexes and demultiplexes control information, together with information in higher layers and error control information, into slot-sized packets. These functions are used to provide three independent services:

- a broadcast service;
- a connection-oriented service;
- a connectionless service.

The broadcast service is a special DECT feature, multiplexing a range of broadcast information into a reserved field. This field appears as part of all active transmissions. The broadcast service is always transmitted in every cell, on at least one physical channel, even in the absence of user traffic. These 'beacon' transmissions allow all portable terminals to quickly identify all fixed terminals that are within range, to select one, and to lock on to it without requiring any mobile transmissions.

9.4.3 The Data Link Layer

The data link layer is concerned with the provision of very reliable data links to the network layer. Many of the imperfections of the radio transmissions are already removed by the effects of the MAC layer, and the DLC is designed to work closely with the MAC layer to provide higher levels of data integrity than can be provided by the MAC layer alone.

The DECT layer model separates into two planes of operation at the DLC layer:

- the control plane, or C-plane;
- the user plane, or U-plane.

The control plane contains all the internal DECT protocol, but may also include some external user information. It always contains protocol entities up to and including the network layer. The C-plane is common to all applications, and provides very reliable links for the transmission of internal control signalling and limited quantities of user information traffic. Full error control is provided with a balanced Link Access Protocol called LAPC.

The user plane contains most of the end-to-end (external) user information and user control. The U-plane provides a family of alternative services, where each service is optimised to the particular need of a specific type of service. The simplest service is the transparent unprotected service used for speech transmission. Other services support circuit mode and packet mode data transmission, with varying levels of protection.

9.4.4 The Network Layer

The network layer is the principal layer involved in the signalling protocol. It is very similar to layer three in the ISDN protocol, and offers similar functionality.

The network layer operates using an exchange of messages between peer entities. This basic message set supports the establishment, maintenance and release of calls. The basic call control provides a circuit switched service selected from one of the range of DLC options. Other network layer services are; supplementary services, the connection-oriented message service, the connectionless message service and mobility management. These services are arranged as independent entities, and a particular application can be realised using more than one.

Mobility management encompasses an important group of services. This group contains the procedures that support the cordless mobility of mobile terminals, for example authentication and location registration.

9.4.5 The Management Entity

The management entity takes care of the following procedures, which concern more than one layer.

MAC layer – the creation, maintenance and release of bearers, by activating and deactivating pairs of physical channels;

DLC layer – the management of connections, which includes the establishment and release of connections in response to network layer demands, the routing of C-plane and U-plane data to appropriate connections.

9.5 DECT Protocols

The embedded protocols are different for the fixed part of the network and the mobile terminal. In Figure 9.4, the software architectures defined in the ETSI standard TER 015 are shown.

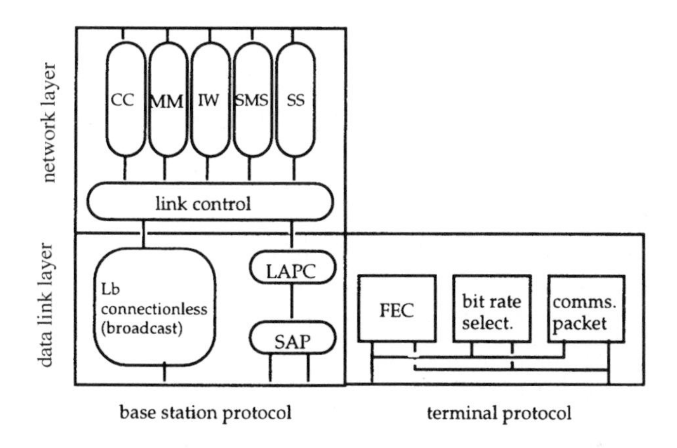

Figure 9.4: Network layer and datalink layer protocols

Two link protocol layers are defined, one for the base station and one for the mobile station.

9.5.1 Base Station Datalink Layer

This layer creates and supervises the base station-to-mobile connection, which transports frames. It uses two protocols: Lb, and Lc + LAPC.

The Lb protocol
The Lb protocol provides a connectionless frame transfer service for broadcasting information. The service is intended mainly for paging messages, and for transmitting alarms.

The Lc + LAPC protocol
The Lc + LAPC protocol provides a connection-oriented service. This protocol is similar to the CCITT recommendation Q.921 for ISDN (LAPD) and

provides a point-to-point service, whereas Lb is used for point-to-multipoint service. The protocol also supports segmentation of long messages.

9.5.2 Mobile Station Datalink Layer

This layer provides basic services for the applications in the higher layers, or for using the terminal as a simple communication bearer.

Frame relay
The mobile datalink layer includes frame relay, which is a real-time packet switching service.

Bit Rate Selection
This service allows applications using a modem to choose the appropriate data transfer speed.

Full Error Correction (FEC)
This service guarantees data transfer without errors for applications requiring high link quality (such as ISDN).

9.5.3 Base Station Network Layer

This layer provides the means of establishing a connection with a mobile station and requesting network services. The standard defines five main functions for this layer:

CC call control
MM mobility management
IW interworking
SS supplementary services
SMS short message service

The higher application layers make use of the services provided by a lower layer through Service Access Points (SAPs), which are the mediators in the client-server relationship.

Call control
The call control function manages the communication link. It negotiates the connection parameters, and establishes, supervises and clears down network connections. The function handles the signalling used to request supplementary services, and offers a significant degree of flexibility.

Mobility management
Mobility management covers the mobility aspect of mobile terminals, and is one of the networks prime functions. It authenticates network users, locates mobile stations in the network, monitors the status of mobiles and identifies mobile stations.

The authentication function uses confidential algorithms and a mechanism based on proof.

The location function enables the mobile station to inform several nodes of its position and its status. This function manages cell handovers of mobile stations, and the channel allocation carried out under the mobile station control as a function of the number of free channels and the received signal quality. Before accessing the network, a mobile station must gather data about the locality in which it is situated, and must check that it is able to establish a connection. In order to achieve this, each base station broadcasts information about the network and its own identity on a radio channel, and in this way a mobile station locates its position and then presents itself to the network.

Interworking unit
The interworking unit enables the suppression of telephony coding during the transfer of data, and takes account of the particular needs of the transfer.

Short message service
The short message function manages the paging service, which enables alphanumeric messages of up to 160 characters to be exchanged (two modes are possible, connection-oriented or connectionless).

Supplementary Services
Supplementary services are functions complementing the standard network services, such as call transfer, charging advice and conditional recall.

9.6 The European View of Next-Generation Cellular Mobile Radio Systems

Second generation cellular mobile radio systems have been widely accepted in Europe. The reasons for the success of GSM are many, one of the most important being that from the beginning of the development, contributions were input from a wide range of experts in several European institutions and companies, leading to widely accepted concepts. A further advantage of GSM is that the standard itself is open and non-proprietary, and thus leaves scope for individual ingenuity from manufacturers and operators, leading in turn to a competitive market. Last, but not least, ongoing deregulation in many European countries has stimulated the mobile communication market and has made a new group of users aware of the convenience of using a mobile phone.

Though second generation mobile systems are just being introduced, a third generation may be just around the corner, for several reasons. GSM is primarily voice-oriented, and the next generation will offer both voice and data services. Research and development activities on the next generation of cellular mobile radio systems are now under way in Europe-the systems are known as Universal Mobile Telecommunications System (UMTS) or International Mobile Telecommunication after the year 2000 (IMT-2000). The third generation cellular mobile radio systems are expected to become operational from the year 2000 and to have the following characteristics:

- high flexibility and spectral efficiency;
- voice communications;
- data communication rate up to 2 Mbps;
- advanced video and data services;
- the incorporation of cordless systems;
- hierarchical cell structures and operation in a wide range of environments;
- the integration of satellite links;
- world-wide roaming and global coverage;
- multi-operator scenarios.

In Europe, work towards a third generation standard is being promoted by a series of activities; for example within the organisation Co-operation in the field of Scientific and Technical research (COST), and the European Union (EU) programme Advanced Communications Technologies and Services (ACTS), which succeeded the RACE programme (Research and Development in Advanced Communications in Europe).

The Organisation of GSM Services

10

10.1 Introduction to the Organisation of GSM Services

The French economy is strongly linked to worldwide exchange and related services occupy a dominant position. Telecommunication networks play a determining role in the efficiency of economic operators:

- due to the capacity of the networks, information, the dominant part of service activities, becomes highly transportable;
- the extent, closeness and interoperation of the networks are the keys to accessibility of information;
- finally, the performance and sophistication of telecommunication networks make them networks for the distribution of services.

In France, the vista of telecommunications has been transformed due to the willingness of the administration to open this branch of the economy to competition, to offer more diversified ranges of products and services and to install new technologies. In this way, with the allocation of licences to the Société Française du Radiotéléphone (SFR) and to Bouygues Télécom, new operators and new areas of activity, such as mobile telecommunications, are introduced.

The French administration now distinguishes two types of operator in the radiotelephone economy: on the one hand the operator and on the other the service provider.

The public authorities choose the operators and give them a concession. In return, the operators have an obligation to offer their services over the whole of the national territory. The aim of this approach is to put operators in competition. Each operator who runs a network constructs, operates and maintains the network according to his objectives.

The service providers form the link between the operator and the mobile telecommunication customers. They create commercial service companies and sell subscriptions to GSM.

10.2 The Operators

There are currently three operators in France, France Télécom, SFR, a subsidiary of Générale des Eaux-the oldest-and the Bouygues group. Allied to the Jean-Claude Decaux group, as well as operating British and American Cable and Wireless and US West, Weba in Germany, the Banque Nationale in Paris and Paribas, the Bouygues group has been assigned a 15 year concession to operate a digital DCS 1800 digital telephone network. This service was commissioned in 1996.

10.3 The Service Marketing Companies

To promote the development of GSM, a new type of distributor has appeared, the service marketing companies. The functions of companies of this type are as follows:

- the sale of GSM terminals;
- the sale of subscriptions to GSM networks;
- the invoicing of subscribers.

A service marketing company is remunerated by operating the traffic generated by its customers on a network. This type of company must widen the GSM market by opening it to new categories of users, by offering them assistance in using the product and made-to-measure services such as:

- explanation of the use of a particular terminal;

- information on the coverage zones of the network;
- access to foreign networks;
- access to services;
- information on the current tariff;
- information on traffic routing;
- assistance in reserving an hotel or restaurant;
- a personalised paging service;
- detailed invoicing.

A service marketing company modifies its tariffs according to defined sections of its customers, for example, according to the average communication time per month or for a particular profession. Similarly, other service providers offer their customers tariffs related to consumption, or hours of usage of the service, per day.

The best known of the service marketing companies are:

- Carrefour;
- CMC (a subsidiary of Matra and Cellcom);
- Locatel services;
- Medès (Debitel France);
- Sagem;
- Sodicam;
- Sodira;
- Vodafone.

Vodafone is the premier European operator of private cellular radio telephones. In Great Britain, Vodafone controls more than 50% of the market with 1.4 million subscribers. In November 1994, Vodafone bought 100% of the shares of Bosch Télécom Service France, thus becoming the leading company in this activity. Furthermore, it should be noted that Vodafone owns SFR shares.

11 The Short Message Service

11.1 Introduction

11.1.1 Definition

The short message service is the facility, offered by a mobile telephone operator to customers, of sending and receiving alphanumeric messages using only a mobile terminal (GSM portable). The maximum message size is 160 characters with a GSM terminal.

A terminal can receive a message on the sole condition of being in service on the network regardless of its state, that is on standby or engaged in communication.

When a terminal sends a message the destination can be either another mobile terminal, a multimedia terminal or even a data terminal.

11.1.2 New Features Introduced with Messages

A short message service provides a simple and effective means of transmitting alphanumeric messages to a GSM terminal or even from a GSM portable to another digital terminal. Hence the GSM terminal is both a digital telephone and a data terminal in a single device. This is new; previously two distinct pieces of apparatus were required.

The prompt message guarantees to the sender that messages have been delivered to the destination, this is the second novelty. A terminal can store the received messages, the service can be extended to receive e-mails. An operator can broadcast information (on the weather, the state of the roads, sporting events, stock market information, etc) to his subscribers.

Unlike previous unilateral paging services Tatoo®, Kobby®, Tam Tam®, the communication is bilateral; a GSM terminal has the capacity to receive and also transmit a message.

Europe is the first continent where a short message service appeared, in 1991 with the introduction of the GSM standard. North America, in contrast, had to wait until 1998 to see the introduction of a comparable service in the digital networks of the operators BellSouth Mobility and Nextel. In North America, this service is part of the new Personal Communication System (PCS) based, like GSM, on TDMA, or even CDMA, multiplexing.

To offer this service, an operator introduces a new component into his network, a message server. This equipment is a repeater, it stores the received message temporarily before delivering it to its destination. Hence, the message source can send when it is ready, independently of the state of the destination receiver. When the destination terminal becomes available for reception, the message server will then transmit the message. A GSM terminal present on the network will receive a message whatever its state, in communication (speech or data transmission) or free (on standby). During periods when a terminal is switched off, it is out of the network and the message server saves messages destined for it in order to retransmit them on its return to the network.

11.2 New Opportunities for Operators

A network equipped with the short message service constitutes a basis for a mobile telephone operator to develop new proposals for value added services. Hence he can differentiate his network from others. The advantages brought by this service are as follows:

- an alterative to previous unilateral alphanumeric paging (without guarantee of delivery to the destination);
- a much greater effective call rate;
- an opening for the introduction of new value added services for subscribers (stock market information, banking services, Virtual Private

Networks (VPN)), and hence new sources of revenue for the operator and support for customer loyalty;

- support for communication between an operator and his customer, to inform him of the cost of his calls, the state of his subscription or even to broadcast information relating to new offers or promotions.

An operator can make use of all these real advantages for modest investment and development costs with regard to the revenue which will be generated, with a marketing campaign targeted at the appropriate category of subscribers who require these services.

The advantages for a subscriber are an increased autonomy with respect to his present environment, a guarantee of confidentiality of his communications with his organisation; the SIM card in his terminal can encrypt if the need is for communication with a Closed User Group (CUG). Reception and reading of a message does not necessarily require isolation of the destination. A terminal can be the communication interface of a microcomputer which needs a modem and a fixed telephone line.

11.3 Architecture of the SMS Network

Figure 11.1 shows the architecture of a GSM network offering the short message service.

11.3.1 The Elements of the Network

A short message centre is an element of the network capable of receiving alphanumeric messages. This element is located in the fixed part of the network.

A short message centre is a data base. It stores a received message before delivering it to the destination terminal. This is a repeater between the calling station and the called station. It introduces asynchronism between transmission and reception of a message. It permits recorded delivery of a message. A message is characterised by a period of validity and a priority. The period of validity determines the duration of storage of a message before its destruction.

The Mobile Switching Centre (MSC) is the interface between the radio subsystem and a wired network. It realises all the required operations for

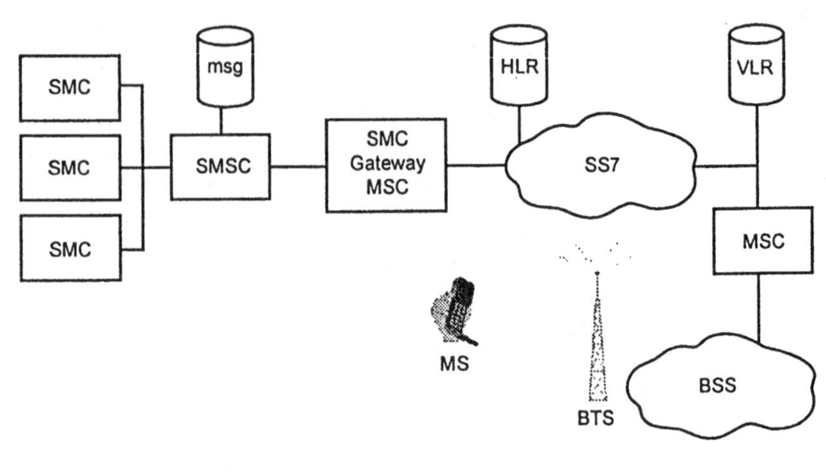

SMC:	Short Message Centre
SMSC	Short Message Service Centre
SMC gateway:	gateway between data and the mobile network (GSM)
HLR:	Home Location Register
VLR:	Visitor Location Register
MSC:	Mobile Switching Centre (GSM switch)
BSS:	Base Station System
MS:	Mobile Station (GSM terminal)

Figure 11.1: Architecture of a GSM network offering the short message service

management of communication with the mobile terminals. To obtain radio coverage of a territory, a mobile network switch controls a group of transmitters. A GSM network with national coverage contains a group of switches.

A Gateway MSC (GMSC) is an interconnecting switch between a mobile network and another network. When a network initiating a call cannot interrogate the nominal data base of a mobile subscriber HLR, it directs the call to a mobile network switch which can itself interrogate the base. The latter consults the data base then routes the call to the destination. The operator of a GSM network alone decides the list of switches which can interrogate the HLRs of his network.

SMS interworking MSC is the interface across which a message sent by a terminal arrives in the short message data base.

SMS gateway MSC is the interface between a short message data base and a GSM network. This interface enables messages to reach the destination terminals.

The functionality of interconnection, InterWorking Function (IWF) is the facility of an MSC switch to interoperate with wired networks, the ISDN, the switched telephone network (PSTN) and data networks. The interconnection characteristics depend greatly on the adjacent network.

The Base Station System (BSS) is the subsystem managing the radio transmitter receiver relays. The MSC switch talks to the subsystem through the interface A. A BSS consists of a station controller BSC and one or more cells and hence one or more base stations BTS.

A BSC station controller manages a group of BTS base stations.

A BTS base station covers one cell of a GSM network.

A mobile station MS is the element of the network used by a subscriber. A mobile terminal, Mobile Equipment (ME) and a SIM together form a mobile station. The mobile terminal divides into two entities, the Mobile Termination, MT and the Terminal Equipment, TE. The characteristics of the mobile termination determine the applications and services supported by a mobile station. The mobile termination performs the following functions:

- the functions of radio transmission and reception;
- management of the radio communication channels;
- support of the man-machine dialogue;
- coding and decoding of speech;
- detection and correction of transmission errors;
- management of signalling with the GSM network;
- management of the transmission speed;
- management of mobility.

The GSM standard defines three types of MT:

- MT0 supports voice and data, without possible extension;
- MT1 is an ISDN terminal with an S bus;
- MT2 is a terminal equipped with a Data inTErface (DTE).

11.3.2 Signalling Within a Network

The arrangement of Figure 11.2 shows the standard architecture of a GSM network interconnected with another network.

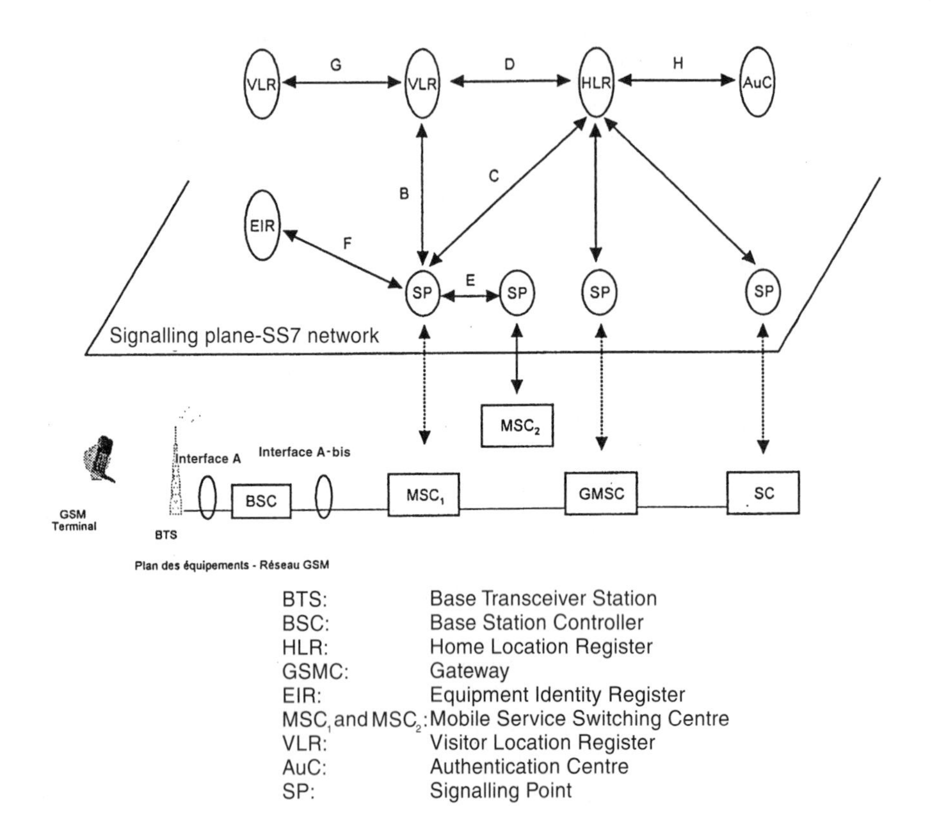

Figure 11.2: Models of the physical architecture and signalling of a GSM network

The signalling scheme of Figure 11.2 shows the signalling interfaces between the functional elements of a network. In this architecture all the functions are distributed among different components-in practice certain functions can be recombined in one piece of equipment. Interfaces B, C, D, E, F and G are built on the Mobile Application Part (MAP) layer of the signalling stack SS7. In contrast, interface H is not standardised.

Interface A, connecting a switch with a BSC, carries commands relating to management of the radio channels, processing of the call and mobility management.

Interface A-bis situated between a BTS and a BSC supports the services offered to users of the network. It serves for frequency management and control of transmitting powers.

Interface B is between a switch and its VLR. It is the data base which stores the information relating to visitors to the network. The switch updates the information in accordance with the activity of visitors entering or leaving the domain as monitored by the switch, or even when a visitor requests use of a particular function.

Interface C is between a switch and its HLR. An interconnecting switch interrogates this base to locate a subscriber and in this way to be able to route a call or message.

Interface D is between an HLR and a VLR. To locate a subscriber it is essential for the operator to keep his spatial coordinates up to date in the HLR base, with the information provided for the VLR.

Interface E is between two switches. When a user crosses the boundary separating the domains of two switches, the switches exchange *handover* messages, control of the subscriber communication passes from one switch to the other. When a short message crosses between a GSM terminal and a short message terminal, interface E intervenes in the transfer of the message, between the MSC switch managing the portable GSM and the MSC managing the short message centre.

Interface F is between an MSC and its EIR. The MSC interrogates the base EIR when it wishes to verify the identity of a mobile station.

Interface G is between two VLRs. This interface is activated when a user leaves the region of a VLR to visit a new domain.

Interface H is between an HLR and an AuC. When an HLR receives a request for subscriber identification or encryption, it interrogates the authentication centre to obtain the information. The GSM standard has not specified the protocol for this interface and consequently it is a proprietary protocol.

11.3.3 Reception of a Message by a Terminal

Figure 11.3 shows the six entities of the network participating in the chain of transmission of a message to a mobile terminal, the types of information and the directions of the exchanges.

- The source of the message (SMC) sends a message to the message server

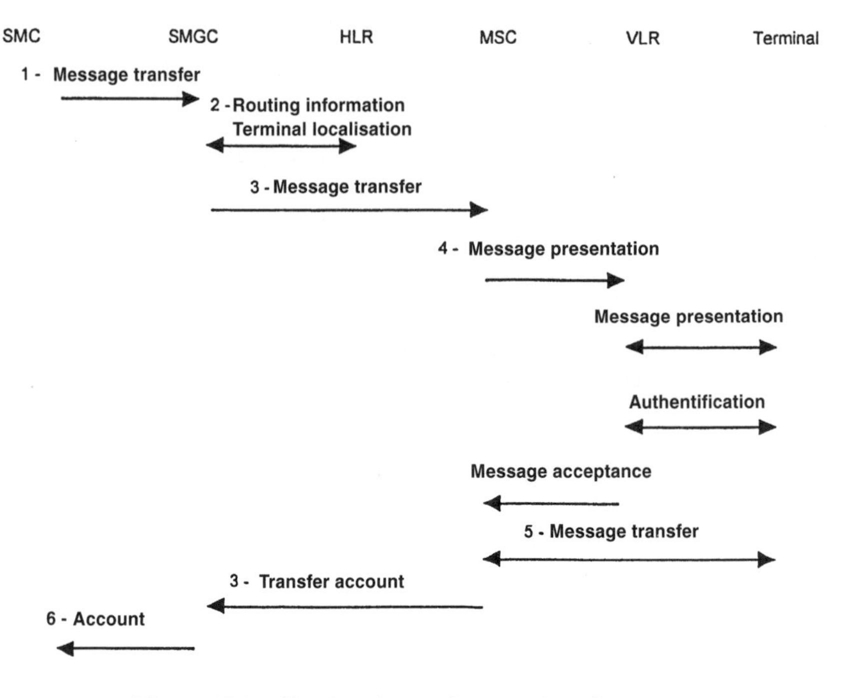

Figure 11.3: Signal exchanges for reception of a message

(SMGC) which records the message and its characteristics (transmitter, destination, priority, limiting date of validity).

- The message server (SMGC) interrogates the HLR of the mobile subscriber to localise it. The HLR provides the references to the switch (MSC) in charge of the mobile subscriber.

- The message server transmits the message and the subscriber references to the switch (MSC) which manages the domain in which the mobile subscriber is located.

- The switch interrogates its visitor data base (VLR) to obtain the most recent coordinates of the subscriber. This request is accompanied by a procedure for message presentation, verification of the state of the terminal, identification and then authentication of the subscriber.

- The switch can finally deliver the message to the terminal.

- The switch provides a detailed account to the message server.

- The message server in turn sends an account to the transmitter.

11.3.4 Transmission of a Message by a Terminal

Figure 11.4 shows the five entities of the network participating in the chain of transmission of a message from a mobile terminal to a data terminal, the types of information and the directions of the exchanges.

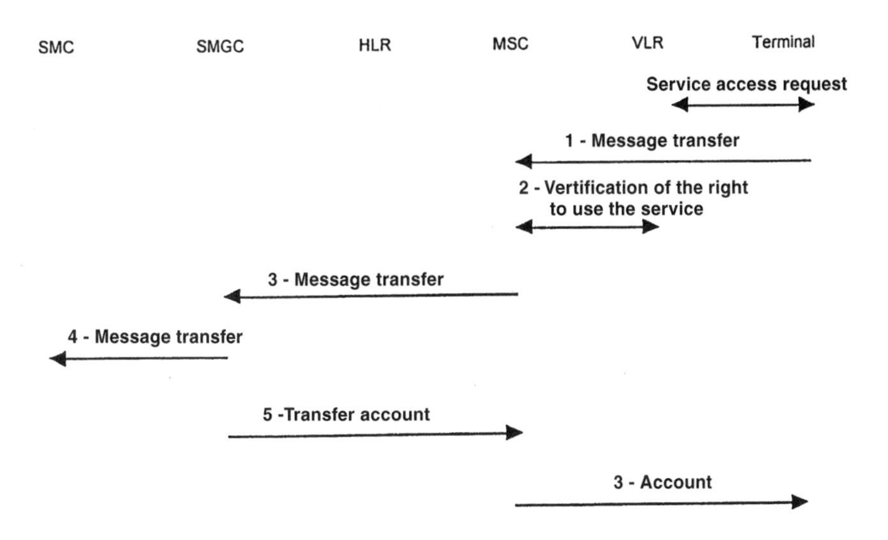

Figure 11.4: Signal exchanges for transmission of a message

- The mobile terminal is the source of the message (MS), it requests authorization to transmit to the VLR. The latter knows the subscriber options and gives authorization.

- The terminal transmits the message to the switch (MSC).

- The switch updates the subscriber register in the VLR, and notes the date and time of the transmission of the message together with the references. The switch can, with this information, deliver an account to the subscriber.

- The switch routes the message to the server centre which stores it and records its characteristics.

- The server retransmits the message to the destination.

- The server provides a detailed account to the switch.

- The switch informs the subscriber of the delivery of the message to the destination.

GPRS

12

12.1 Introduction

GSM networks offer a data transmission service but the rate is limited to a maximum of 9.6 kbit/s. GSM networks are based on circuit switching. A communication channel is occupied by a user during a communication. This mode of transmission is suitable for voice transport. But the future of GSM now extends to offering data services. GSM operators wish to specify better quality and more competitive data services in order to respond to customer expectations. It is for this reason that European Telecommunications Standards Institute (ETSI) recommends integration of the techniques of transmission 'by packets'; for this ETSI has published GSM Phase 2+ specifications, with the purpose of introducing a new technology called General Packet Radio Service (GPRS) into GSM networks.

GPRS permits access to Internet services with an effective data rate up to 115 kbit/s by using multiple radio channels (up to eight) which are allocated to one user or shared by several users. With GPRS, the radio resources are allocated dynamically and the transmission rate varies due to greater flexibility and greater adaptability of the packet mode with respect to the circuit mode. GPRS uses statistical multiplexing.

Mobile Internet or intranet services made available by the deployment of GPRS are the mobile office (*remote access* or connection at a distance into the organisation's network), electronic mail, Internet access, electronic commerce, localised information services and telemetry.

In fact, GPRS is particularly efficient for transmission of discontinuous data or frequent transmission of small volumes of data. Media applications of GPRS will permit rapid access to directory services such as Yellow Pages, remote on line loading of audio files, etc.

Figure 12.1 shows the architecture and interfaces between a GSM-GPRS network providing a packet mode service and a packet mode transmission network. When a GPRS network is interconnected with an IP network, it behaves like an IP network; the point Gi of Figure 12.1 is the interface between the two data networks.

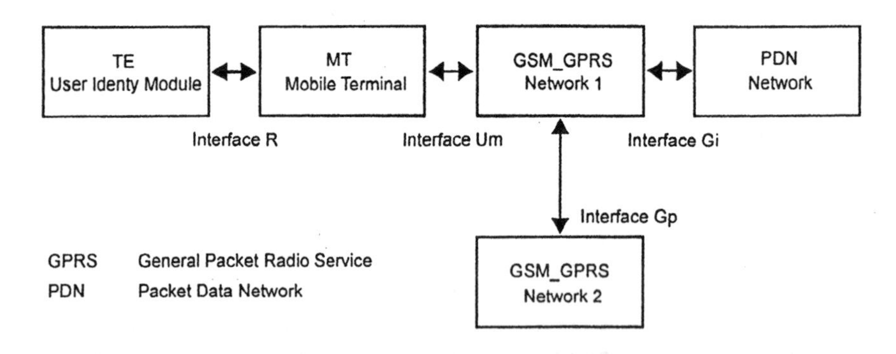

Figure 12.1: Architecture of a GPRS system with reference and interface points

Generally two IP networks interface through a router. The reference point Gi is the boundary between a Gateway GPRS Support Node (GGSN) and an external IP network. For the external IP network, the GGSN is seen as a router. Layers L1 and L2 of the software stacks (Figure 12.2) are specific to the operator of the GSM-GPRS network.

GPRS is a major evolution of GSM which requires the introduction of new elements into the network. Hence GSM offers circuit mode services for speech and packet mode services for data.

12.2 The Components of a GPRS Network

Figure 12.2 shows the architecture of the software stacks in each of the elements of a GPRS network.

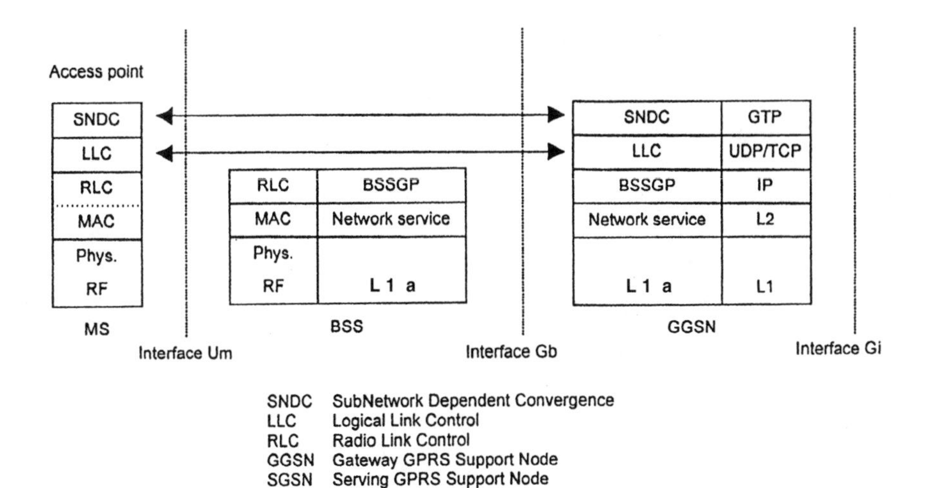

Figure 12.2: The software stacks of a GPRS system

From bottom to top the following software occurs in the 'MS' terminal:

- The physical layer, which divides into two functional sub-layers;

- The RF sub-layer manages the radio functions of the terminal. It transmits the information received from the physical layer. It decodes the radio signals received from the base station and transfers them for interpretation to the physical layer.

- The physical layer produces the frames which will be transmitted by the radio layer; it detects and corrects transmission errors in frames received from the network.

- The MAC layer or Radio Link Control (RLC) drives the radio link between the terminal and the BTS, that is the retransmission mechanisms in case of error and the radio resource access control function when several terminals are in competition. The RLC can request retransmission of a block of data.

- The upper layer SubNetwork Dependant Convergence (SNDC) manages the movement, encoding and compression of data.

New GPRS compatible terminals are necessary to benefit from the new GPRS services.

The GGSN performs the interconnection function in the MSC, which permits communication with other packet data networks external to the GSM network. The GGSN conceals the specific details of GPRS from the data network. The GGSN manages the invoicing of subscribers to the service. The GGSN must support the protocol used on the data network with which it is interconnected. The data protocols supported as standard by a GGSN are IP v6, CLNP and X25.

The Serving GPRS Support Node (SGSN) is the service function in the MSC which permits management of the services offered to the user. The SGSN is the logical interface between the GSM subscriber and an external data network. The principal functions of the SGSN are, on the one hand management of active mobile subscribers (continuous updating of the references of a subscriber and the services used) and on the other hand relaying of data packets. When a data packet arrives from a Packet Data Network (PDN) external to the GSM network, the GGSN receives this packet and transfers it to the SGSN which retransmits it to the mobile station; for the outgoing packets it is the SGN which transits them to the GGSN.

12.2.1 Connection in Transparent Mode

A GSM network must offer connections to the Internet and to intranet networks. These networks have specific requirements such as subscriber identification, authentication, ciphering of data and even dynamic allocation of the IP address between a GSM terminal and an intranet network. A GPRS network has two response modes to this, either by providing a transparent mode or by providing the functions requested by the intranet network. Figure 12.3 shows the software components.

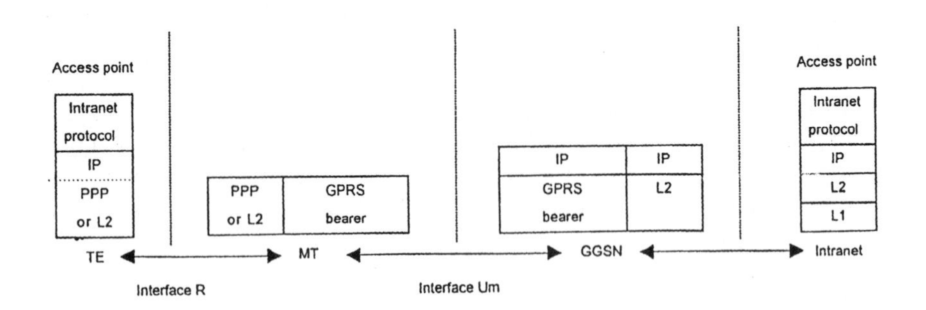

Figure 12.3: Transparent access through an IP network (the software stacks)

The GSM terminal (TE + MT) has an IP address within the address range of the operator. The address is either a static public address allocated with the subscription, or a dynamic address provided during connection. This address serves for the transfer of data between the terminal and the intranet network.

Figure 12.4 shows the connection of a portable PC and a GSM terminal.

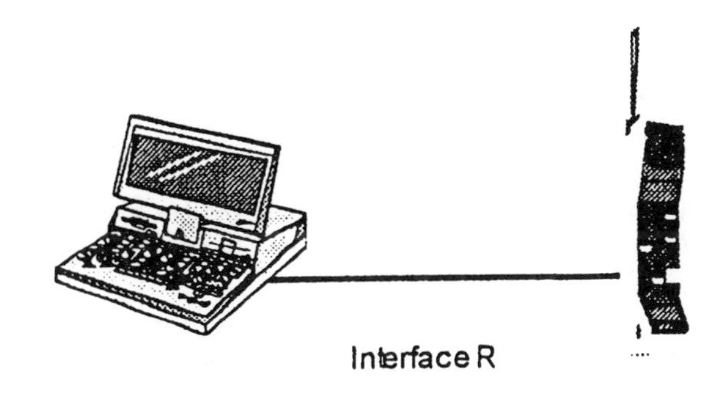

Figure 12.4: Interface R between a PC and a GSM terminal

12.2.2 Connection in Non-Transparent Mode

The GSM terminal (TE + MT) has a private IP address within the address range of the Internet Service Provider (ISP) or even the intranet.

12.2.3 Packet Routing

Routing of each packet is independent of that of its predecessor or that which follows it. During the connection phase of a terminal on a GSM network, there are numerous exchanges of signals, and to meet the constraints of the packet mode, the routing information to direct the first packet to a GSM terminal is stored in the GGSN. Hence the route for the following packets is selected from the context stored in the GGSN, Temporary Logical Link Identity (TLLI).

12.3 The WAP Protocol

The Wireless Application Protocol (WAP) is a new protocol intended for multimedia and Internet applications relying on GSM. In fact a portable is henceforth a device which is provided with a personal configuration by its user; it also permits identification of a user by a service provider.

The WAP protocol defines an application environment and a communication protocol.

Two elements make up the application environment: a WML language which permits definition of an interface for the user independent of the support and a programming language WML Script; in this way the application designers can create the application. This amounts to providing a *micro-browser* for the GSM terminal user similar to the *browser* used to access the Web.

The communication protocol has a two layer structure. Figure 12.5 shows the Internet and WAP software stacks.

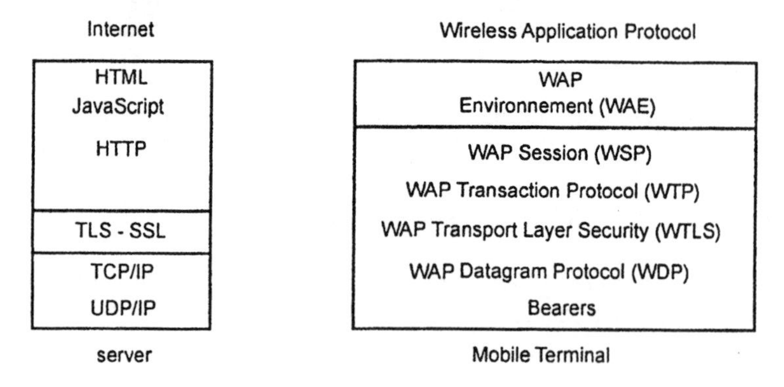

Figure 12.5: Internet and WAP software stacks

The WAP software stack is independent of the transport service supporting it (the *bearer*), hence an application can operate with different transport services.

12.4 The Future of GPRS

The target of GPRS is the community of Internet users and electronic mail. GPRS introduces a new support for already widely extended services and it responds to the need for user mobility. One possible scenario for the evolution of GPRS is as follows:

- Phase 1-introduction of the service: existing GPRS terminals offer few functions, the operators must attract the market with services and competitive costs.
- Phase 2-expansion of the service: the number of services increases, term-

inals have new functions, the network permits a higher rate of transfer.

- Phase 3-GPRS becomes widespread, the rate of penetration of the service is high, competition between operators and service providers intensifies.

12.5 The Advantages of GPRS

For operators, GPRS has numerous assets:

- GSM operators profit from the surge of the Internet to attract users and increase their activity;
- the operators optimise the transport of data in their network;
- the operators can introduce new innovative services;
- mobile networks can compete increasingly with fixed networks, the differential between the families of networks becomes blurred;
- GSM networks evolve progressively towards the third generation.

For users, the attractions of GPRS are:

- access from a GSM terminal to Internet and intranet networks;
- all the applications available on a LAN are accessible from a cordless terminal. GPRS eliminates the physical link between a PC and a LAN;
- invoicing by usage.

UMTS

13

13.1 Introduction

Working group 8/1 of the International Telecommunications Union (ITU) is currently working on one its most ambitious projects-standardisation of the third generation mobile telecommunication systems which provide radio access to the world telecommunications infrastructure, on a worldwide basis, by including both satellite systems and terrestrial methods serving fixed and mobile users of public and private networks. Called International Mobile Telecommunication System 2000 (IMT-2000), this project will permit every user to join regardless of location and time by ensuring transparent operation of mobile terminals throughout the world.

The system definition covers various concepts, including the service plan, the radio infrastructures and the network infrastructures. Development of these concepts relies on the one hand on a predictive analysis of the evolution of the services and networks, mobile and fixed, and on the other hand on the prospects of development of new techniques which will permit extending the operational limits currently encountered. The objectives of project IMT-2000 are:

- Definition of a framework so that the systems of this family are compatible with each other.
- A cellular system compatible with second generation systems.

- Speech services.
- Data services.
- Standardised components and interfaces.
- A philosophy of universal wireless communication, for which the networks would be capable of establishing communication not only between fixed terminals, but also between persons regardless of location and movement. A subscriber would have a multifunction personal pocket terminal, for use at home, at the office, while travelling by car or on foot, in public places, in vehicles and so on.

Figure 13.1 shows the components and interfaces of the future IMT-2000 standard.

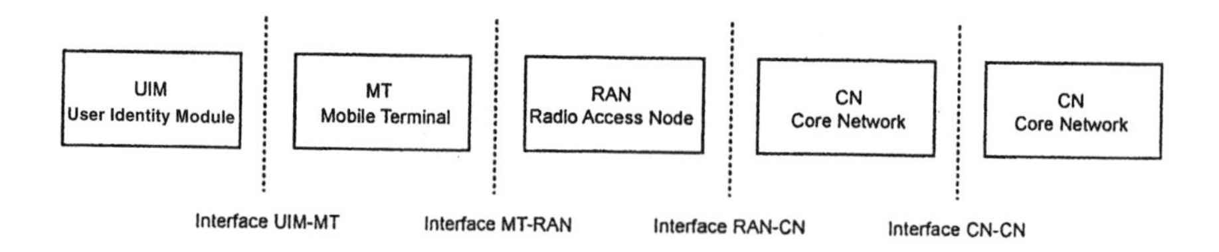

Figure 13.1: Architecture of an IMT-2000 system.

The four major components are:

- User Identity Module (UIM): this is the concept of the SIM card of GSM, thus a subscriber can use different terminals according to the circumstances.
- Mobile Termina (MT): the mobile terminal.
- Radio Access Network (RAN): this is the means of access to the network, it varies according to the environment, radio station or satellite.
- Core Network (CN): this is the network and its functions.

In the context of the ITU work for the IMT-2000 project, ETSI is working to produce a European standard, it is the UMTS project which will be a member of the IMT-2000 family.

UMTS networks will constitute mobile and wireless third generation mobile telecommunication systems, capable of offering high data rate services of the multimedia type to the general public. The definition and objectives assigned to the project by the European Parliament are as follows.

Definition
By UMTS the European Parliament means a mobile communication system of the third generation capable of providing, in particular, wireless multimedia services of a new type exceeding the current possibilities of second generation systems, such as GSM, and combining the use of terrestrial and satellite elements.

UMTS developed in the Community must be compatible with the concept of the third generation system IMT-2000, which has been drafted by the ITU at world level on the basis of ITU Resolution 212.

The European Parliament defined the following characteristics for UMTS networks.

For the services

- Multimedia and mobility capabilities over a very large geographical extent.
- Efficient access to the Internet, intranets and other services based on the Internet Protocol (IP).
- High quality voice transmission, comparable to that of fixed networks.
- Portability of services in different UMTS environments.
- Operation in GSM/UMTS mode inside and outside and at distant external locations, without a break, permitting free movement between GSM networks and between terrestrial and satellite elements of UMTS networks.

For the terminals

- Bimodal GSM/UMTS terminals on two bands if appropriate.
- Bimodal terrestrial/satellite UMTS terminals if appropriate.

The radio access interface
Working group 8/1 of the ITU is currently examining 16 proposals for radio interfaces (ten terrestrial and six satellite) for third generation systems, of which that of UMTS Terrestrial Radio Access (UTRA) is defined by ETSI. It is continuing its work to define the principal characteristics of IMT-2000 radio interfaces.

Third generation mobile services will ensure access, by means of one or more radio links, to numerous services offered by fixed networks, the Internet and other specific mobile services. Various types of mobile terminal, for fixed or mobile use, will be developed with links to terrestrial services and/or by satellite.

The basic network

- Development from GSM systems; management of the mobility of call commands consisting of a total routing function based on the basic standard of the GSM network.

- Elements of convergence mobile/fixed.

13.2 The International Context

Future UMTS systems will exploit the frequency bands reserved by the World Radiocommunications Conference in 1992, with the designation IMT-2000.

On the 14th of December 1998 the cabinet and the European Parliament accepted a resolution relating to the introduction of UMTS on its territory. In article three, this resolution states that:

> Member States will take all necessary measures to permit harmonised provision of UMTS services on their territory by the 1st of January 2002 at the latest, and will establish a system of authorization for UMTS by the 1st of January 2000 at the latest.
>
> In the preparation and application of their authorization regimes, the Member States will ensure that the provision of UMTS services is organised in the frequency bands which are harmonised by CEPT, and with respect to the European standards formulated by ETSI where these exist, including in particular a common aerial interface standard which is open and competitive within the international plan. Member States will ensure that licences permit travel throughout the community territory.

13.3 The French Context

The telecommunications regulation law of 26th July 1996 describes the case where the number of authorisations can be limited. It specifies the respective powers of the various administrative bodies in charge of the telecommunications sector in this case.

In particular, article L.33-1 V of the Code of Posts and Telecommunications permits:

> The number of authorisations can be limited as a consequence of technical constraints inherent in the availability of frequencies.
>
> In this case, the minister responsible for telecommunications, in a proposal of the telecommunications regulating authority, publishes the terms and conditions of allocation of authorisations.

13.4 The UMTS Stakes

13.4.1 Services

The services proposed by the companies have two components: the supply of information (the contents) and the technical support which carries it that is the container. The market, if there is one, will be formed from a combination of this supply with user demand.

Today, the principal activity of GSM operators is the provision of the voice telephone service. The contents carried by the networks are thus created by the users themselves. The development of advanced services, of the multimedia type, on UMTS networks, implies the existence of content providers and raises the role of service providers, responsible for setting up 'embellishments' at their customers destination.

Quality of service: the GSM standard offers a single level of service quality matched to the telephone service. For UMTS several levels of service quality are desired.

The requirements of applications in terms of bit rate, response time and error rate are different, some applications require a constant response time, others are less sensitive to error rate. For example an e-mail has little sensitivity to transit time; in contrast, an interactive application such as a game requires the commands to traverse the network rapidly, although a command contains little data, perhaps 10 bytes that is 80 bits. The transfer capacity of GSM is under utilised with an 80/9600 game that represents 0.8% of a circuit; in contrast the radio interface resources are better utilised in packet mode with GPRS or UMTS.

13.4.2 The Technological Stakes

Standardisation

Standardisation will be a key factor in offering quality services at reasonable cost and to permit movement between systems; its success depends on the flexibility of the interfaces and the capacity to evolve in parallel with the technology. Continuous and tight cooperation between operators, equipment suppliers and regulators in the standardisation of UMTS is crucial for UMTS to achieve the same success as GSM.

ETSI is working on the definition of the UMTS standard. The latter must succeed the GSM standard. The fixed objective is to permit wide band communication through a single interface and multimedia services will be accessible from a mobile terminal (interactive video, data transfer, Internet, etc.).

Radio interface

The GSM standard is now evolving both to increase the bit rate and to introduce new modes: High Speed Circuit Switched Data (HSCSD) and/or General Packet Radio Services (GPRS). But for the future UMTS system, the terrestrial radio interface is based on a new radio interface-UTRA-distinct from that of the GSM system. ETSI Working group SMG2 is working to provide a proposed standard for the end of 1999. The proposals submitted to ETSI for this radio interface rely on two multiplexing techniques: TDMA (TDD) and Wideband Code Division Multiple Access (W-CDMA (FDD)). The UTRA proposal will be submitted to the ITU for project IMT-2000.

The interface data rate depends on the user environment:

- in rural areas: at least 144 kbit/s, the aim is 384 kbit/s;
- in urban areas: at least 384 kbit/s, the aim is 512 kbit/s;
- in a building: at least 2 Mbit/s.

The UTRA interface must offer negotiation of the attributes of services (type of support, data rate, error rate, end to end transmission delay, etc.), support for circuit and packet oriented services, priority management on the radio interface, matching of the link to the quality and load on the network.

The UTRA interface must offer *handover* without cutoff from the network of one UMTS operator to that of another UMTS operator, and also to a second generation GSM network.

Networks

UMTS networks must respond after the year 2000 to a mass market for communication services with mobiles. Prospective studies predict 50 to 100 million European subscribers in 2000, and 100 to 200 million in 2010. This indicates that the traffic carried by mobile terminals will approach that carried by wired terminals.

A second objective of UMTS is to offer a universally mobile system, surpassing the limitations due to the multiplicity of systems and networks. Hence a single terminal will permit communication in numerous user environments: home, office, street, car, train, etc. The terminal must manage a large diversity of radio coverage, from short range radio terminals in buildings to satellite spots covering areas of low population density for example. The coverage of the networks will be worldwide, but mobility must be adapted to the requirements corresponding to the lifestyle of each user.

Another objective of UMTS is to support new high speed services. The increasing requirements of data and video services (database, file transfer, high definition facsimile, mobile visionphone, etc.), and the interconnection to wideband networks and ISDN networks, motivates a diversification of terminals and the increase of data rates available at the radio interface. The limit could be around 2 Mbit/s in some user environments. The quality of service must be equivalent to that of wired networks.

Terminals

The integration of operating systems and applications within the UMTS terminal marks the convergence of computer and communication technologies.

The limits related to the diversity of systems could be partially avoided to offer a personal communication service, either by realising multimode terminals (GSM-DECT, GSM-satellite, etc.) or by proposing subscriber cards (SIM cards) usable for several terminals and infrastructures.

A terminal must be capable of operating in four types of environment: with a satellite, in a rural area, in an urban area and in a building. Figure 13.2 shows the environments and terminal families.

Security

A smart card ensures the security of the terminal and the confidentiality of communication, encrypting algorithms with public keys are used.

The Advanced Security for Personal Communication Technologies (**ASPeCT**) project has evaluated the solutions to the principal problems of security while using UMTS technologies. The identification tests conducted by ASPeCT show the benefits of an infrastructure with a public key for identification of users of the network. Payment methods for secure invoicing

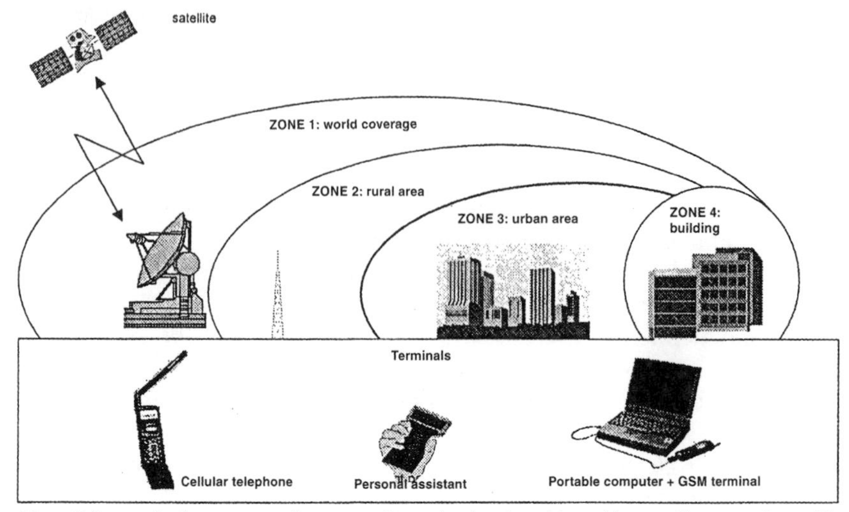

The cellular terminal can communicate according to its situation either with a satellite network or with a GSM network. UMTS defines the coverage domain as follows: coverage by a constellation of satellites is worldwide; coverage by a GSM network is national; a GSM macrocell covers a suburban area; a GSM microcell covers an urban area; a GSM picocell covers the interior of a building

Figure 13.2: Architecture of a UMTS network

have been demonstrated with a mobile user, a value added service provider and a certification authority. The partners of ASPeCT already use the demonstration of secure invoicing in their electronic commerce activity. The UMTS SECurity Architectures (**USECA**) project defines a set of mechanisms, protocols and security procedures together with a functional and physical architecture for security. A number of possibilities described as third generation will be included: detection of false base stations, limitation of the transmission of encoding keys of identification value between networks, use of longer keys, independent identification of encoding, protection of the integrity of data and protection of the identity of the terminal.

Access to online services

Access to services on line from a mobile responds to specific requirements. Developments are in progress in the GSM(WAP[1]) domain with the introduction of a navigator adapted to the mobile environment.

[1] Wireless Application Protocol.

The WAP Protocol

14

14.1 The Context

14.1.1 What is WAP?

The Wireless Application Protocol (WAP) bridges the gap between the mobile world and the Internet as well as corporate intranets and offers the ability to deliver an unlimited range of mobile value-added services to subscribers – independent of their network, bearer, and terminal. Mobile subscribers can access the same wealth of information from a pocket-sized device as they can from the desktop.

WAP defines an application environment Wireless Application Environment (WAE) aimed at enabling operators, manufacturers, and content developers to develop advanced differentiating services and applications including a microbrowser, scripting facilities, e-mail, World Wide Web (WWW) to mobile-handset messaging, and mobile to telefax access.

14.1.2 What is the Purpose of WAP?

WAP has the following goals:

- To bring Internet content and advanced data services to wireless phones and other wireless terminals.

- To create a global wireless protocol specification that works across all wireless network technologies.
- To enable the creation of content and applications that cover a wide range of wireless bearer networks and device types.
- To embrace and extend existing standards and technology wherever possible and appropriate.
- To enable fast easy delivery of relevant information and services to mobile users.

WAP makes minimal demands on the air interface itself. The WAP specification can operate on the widest number of air interfaces. It defines a protocol stack that can operate on high latency, low bandwidth networks such as Short Message Service (SMS), or GSM Unstructured Supplementary Service Data (USSD) channel. The WAP specification is also independent of any particular device. Instead, it specifies the bare minimum functionality a device must have, and has been designed to accommodate any functionality above that minimum.

Bearer and device independence both help to foster interoperability. But interoperability goes beyond these two principles to require that each WAP-compatible component will communicate with all other components in the proposed network by using the standard methods and protocols defined in the specification.

Interoperability provides clear benefits for handset manufacturers and infrastructure providers. Handset manufacturers are assured that if their device complies with the WAP specification it will be able to interface with any WAP-compliant server, regardless of the manufacturer. Likewise, the makers of a WAP-compliant server are assured that any WAP-compliant handset will interface correctly with their servers.

14.1.3 What is the WAP Forum?

Ericsson, Nokia, Motorola, and Unwired Planet founded the WAP Forum in the summer of 1997 with the purpose of defining an industry-wide specification for developing applications over wireless communication networks. The WAP specifications define a set of protocols, in the application, session, transaction, security, and transport layers, which enable operators, manufacturers and application providers to meet the challenges of advanced wireless service differentiation and fast/flexible service creation. There are now over 200 members representing terminal and infrastructure manufacturers, operators, carriers, service providers, software houses, content providers and companies developing services and applications for mobile devices.

The WAP Forum currently has several different relationships with other standards bodies:

- The WAP Forum is submitting its specifications to the European Telecommunications Standards Institute (ETSI). In addition to having a formal liaison between the two groups, the Mobile Execution Environment (MExE) subgroup within ETSI's Special Mobile Group four is cross-referencing the WAP specification to define a compliance profile for GSM and UMTS.

- The Cellular Telecommunications Industry Association (CTIA) has an official Liaison Officer to the WAP Forum.

- The WAP Forum has established a formal liaison relationship with the World Wide Web Consortium (W3C) and the Telecommunications Industry Association (TIA). The WAP Forum is collaborating with these organisations in the area of WWW technologies in the wireless sector. The W3C, TIA and the WAP Forum intend to continue to work together in selected technical areas to jointly create and promote technical specifications of interest to all three organisations.

- The WAP Forum is in the process of forming a liaison relationship with the Internet Engineering Task Force (IETF).

14.1.4 Which Wireless Networks for the WAP?

Wireless data networks present a more constrained communication environment than wired networks. Because of fundamental limitations of power, available spectrum and mobility, wireless data networks tend to have:

- Less bandwidth

- More latency

- Less connection stability

- Less predictable availability

Furthermore, as bandwidth increases, the handset's power consumption also increases which further taxes the already limited battery life of a mobile device. Therefore, even as wireless networks improve their ability to deliver higher bandwidth, the power availability at the handset will still limit the effective throughput of data to and from the device. A wireless data system must be able to overcome these network limitations and still deliver a satisfactory user experience.

WAP is designed to work with most wireless networks such as Cellular Digital Packet Data (CDPD), CDMA, GSM, PDC, PHS[1], TDMA, iDEN[2], TETRA[3], DECT.

14.1.5 The Benefits

Operators

For wireless network operators, WAP promises to decrease churn, cut costs, and increase the subscriber base both by improving existing services, such as interfaces to voice mail and prepaid systems, and facilitating an unlimited range of new value-added services and applications, such as account management and billing inquiries. New applications can be introduced quickly and easily without the need for additional infrastructure or modifications to the phone. This will allow operators to differentiate themselves from their competitors with new, customised information services. WAP is an interoperable framework, enabling the provision of end to end turnkey solutions that will create a lasting competitive advantage, build consumer loyalty, and increase revenues.

Content providers

Applications will be written in Wireless Markup Language (WML), which is a subset of eXtensible Markup Language (XML). Using the same model as the Internet, WAP will enable content and application developers to grasp the tag-based WML that will pave the way for services to be written and deployed within an operator's network quickly and easily. As WAP is a global and interoperable open standard, content providers have immediate access to a wealth of potential customers who will seek such applications to enhance the service offerings given to their own existing and potential subscriber base. Mobile consumers are becoming more hungry for increased functionality and added value from their mobile devices, and WAP opens the door to this untapped market that is expected to reach 100 million WAP-enabled devices by the end of 2000. This presents developers with significant revenue opportunities.

[1] Personal Handyphone System: Japanese digital cordless standard; uses very small cells, 1900-MHz band, and data rates to 64 kbit/s when stationary.

[2] Integrated Digital Enhanced Network: Proprietary digital TDMA system developed by Motorola that supports both packet and circuit switching and offers each user up to six 10-kbit/s channels.

[3] Terrestrial Trunked Radio: ETSI standard for packet-switched digital radio networks that allows terminals to communicate directly where there is no cellular coverage.

End users

End users of WAP will benefit from easy, secure access to relevant Internet information and services such as unified messaging, banking, and entertainment through their mobile devices. Intranet information such as corporate databases can also be accessed via WAP technology. Because a wide range of handset manufacturers already supports the WAP initiative, users will have significant freedom of choice when selecting mobile terminals and the applications they support. Users will be able to receive and request information in a controlled, fast, and low-cost environment, a fact that renders WAP services more attractive to consumers who demand more value and functionality from their mobile terminals.

As the initial focus of WAP, the Internet will set many of the trends in advance of WAP implementation. It is expected that the ISPs will exploit the true potential of WAP. Web content developers will have great knowledge and direct access to the people they attempt to reach. In addition, these developers will be likely to acknowledge the huge potential of the operator's customer bases; thus, they will be willing and able to offer competitive prices for their content. WAP's push capability will enable weather and travel information providers to use WAP. This push mechanism affords a distinct advantage over the WWW and represents tremendous potential for both information providers and mobile operators.

14.2 The Wireless Application Environment

The WAP Forum produces specifications. One of these documents is the Wireless Application Environment (WAE). This document defines an application framework for wireless devices such as mobile telephones, pagers and PDAs. The objectives of the WAE effort are:

- To define an application architecture model which is suitable for building interactive applications that function well on devices with limited capabilities and in narrow-band environments with high latencies.
- To define a general-purpose application programming model.
- To provide Network Operators with the means to enhance and extend network services.
- To enable multi-vendor interoperability.
- Not to dictate or assume any particular Man-Machine-Interface (MMI) model.

- To provide the means to communicate device capabilities to origin servers.

14.2.1 The WAE Architecture Model

The WWW architecture model

The architecture model includes all elements of the WAP architecture related to application, specification and execution.

The architecture defines primarily the network elements, content formats, and shared services. The Interfaces are not standardised and are specific to a particular implementation.

The architecture model is based upon the Client–Server logical model and Figure 14.1 shows this architecture.

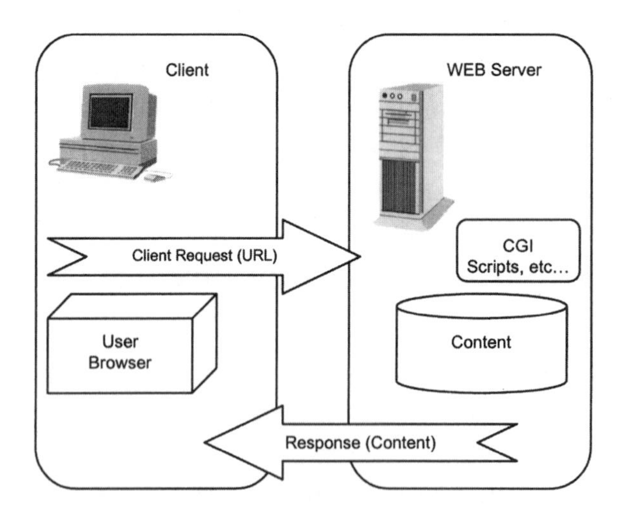

Figure 14.1: WEB model

The Internet's World Wide Web (WWW) uses the Client–Server model, which is a flexible and powerful logical model. The Servers (Applications) present content to a client in a set of standard data formats that are *browsed* by client-side user agents known as Web browsers. A Client (user agent) sends requests for one or more named data objects (or content) to a server. The server responds with the requested data expressed in one of the standard formats known to the Client.

This infrastructure allows users to easily reach a large number of third-party applications and content services. It also allows application developers to easily create applications and content services for a large community of clients.

The WWW standards include all the mechanisms necessary to build a general-purpose environment, including:

- Standard naming model: all servers and content on the WWW are named with an Internet-standard *Uniform Resource Locator*(URL) [RFC1738, RFC1808].

- Content typing: all content on the WWW is given a specific type thereby allowing web browsers to correctly process the content based on its type [RFC2045, RFC2048].

- Standard content formats: all web browsers support a set of standard content formats. These include the HyperText Markup Language (HTML), the JavaScript scripting language [ECMAScript JavaScript].

- Standard Protocols: standard networking protocols allow any web browser to communicate with any web server. The most commonly used protocol on the WWW is the HyperText Transport Protocol (HTTP) [RFC2068].

The WWW protocols define three classes of server:

1 Origin server: the server on which a given resource (content) resides or is to be created.
2 Proxy: an intermediary program that acts as both a server and a client for the purpose of making requests on behalf of other clients. The proxy typically resides between clients and servers that have no means of direct communication, e.g. across a firewall. Requests are either serviced by the proxy program or passed on, with possible translation, to other servers. A proxy must implement both the client and server requirements of the WWW specifications.
3 Gateway: a server which acts as an intermediary for some other server. Unlike a proxy, a gateway receives requests as if it were the origin server for the requested resource. The requesting client may not be aware that it is communicating with a gateway.

The WAE architecture model

WAE adopts a model that closely follows the WWW model. All content is specified in formats that are similar to the standard Internet formats. Content is transported using standard protocols in the WWW domain and an optimised HTTP-like protocol in the wireless domain.

WAE has borrowed standards from WWW including authoring and publishing methods wherever possible. The WAE architecture allows all content and services to be hosted on standard Web origin servers that can incorporate proven technologies (e.g. CGI). All content is located using WWW standard URLs.

WAE enhances some of the WWW standards in ways that reflect the device and network characteristics. WAE extensions are added to support Mobile Network Services such as Call Control and Messaging. Support for low bandwidth and high latency networks is also included in the architecture.

WAE assumes the existence of *gateway* functionality responsible for encoding and decoding data transferred from and to the mobile client. The purpose of encoding content delivered to the client is to minimise the size of data sent to the client over-the-air as well as to minimise the computational energy required by the client to process that data. The gateway functionality can be added to origin servers or placed in dedicated gateways as illustrated in Figure 14.2.

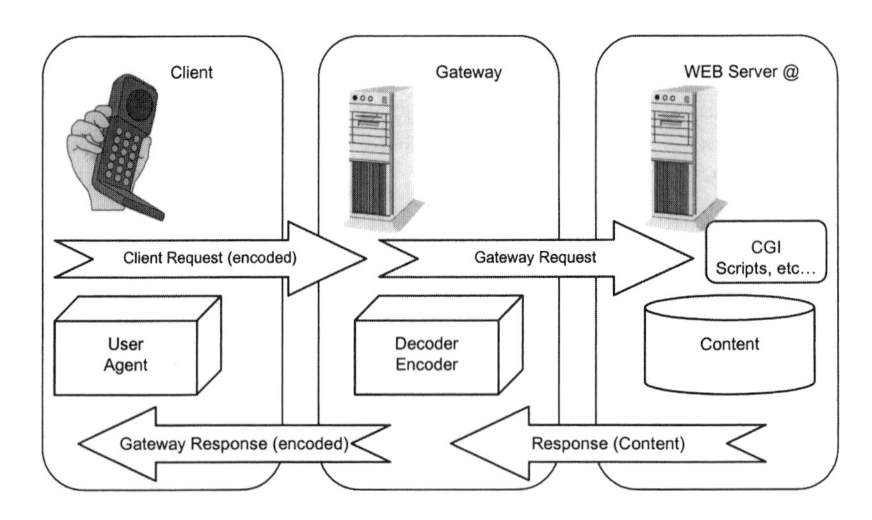

Figure 14.2: WAP environment model

The WAP environment model includes the following elements:

1 User Agent: client-side in-device software that provides specific functionality (e.g. display content) to the end-user. User Agent interprets network content. WAE includes user agents for the two primary standard contents: encoded WML and compiled Wireless Markup Language Script (WMLScript).
2 Server (Content Generators): applications (or services) on origin servers

(e.g. CGI scripts) that produce standard content formats in response to requests from user agents in the mobile terminal. WAE does not specify any standard content generators.

3 Gateway (Standard Content Encoding): the WAP gateway provides the handset with computing resources. Well-defined content encoding allows a WAE user agent (e.g. a browser) to conveniently navigate web content. Standard content encoding includes compressed encoding for WML, bytecode encoding for WMLScript, standard image formats, a multi-part container format and adopted business and calendar data formats. A WAP Gateway will typically take over all DNS services to resolve domain names used in URLs, thus offloading this computing task from the handset. The WAP Gateway can also be used to provide services to subscribers and provide the network operator with a control point to manage fraud and service utilisation.
4 Wireless Telephony Applications (WTA): a collection of telephony specific extensions for call and feature control mechanisms that provide authors (and ultimately end-users) with advanced Mobile Network Services.

Typically, a user agent on the terminal initiates a request for content. However, not all content delivered to the terminal will result from a terminal-side request. For example, WTA includes mechanisms that allow origin servers to deliver generated content to the terminal without a terminal's request.

In some cases, what the origin server delivers to the device may depend on the characteristics of the device. The Client characteristics are communicated to the server via standard capability negotiation mechanisms that allows applications on the origin server to determine the characteristics of the mobile terminal device. WAE defines a set of user agent capabilities that will be exchanged using Wireless Session Protocol (WSP) mechanisms. These capabilities include such global device characteristics as WML version supported, WMLScript version supported, floating-point support, image formats supported and so on.

WAE is based on the architecture used for WWW proxy servers. The situation where a user agent must connect through a proxy to reach an origin server (i.e., the server that contains the desired content) is very similar to the case of a wireless device accessing a server through a gateway. Figure 14.3 shows the Wireless Application Environment network components.

The connections between the user agent and the gateway use WSP, regardless of the protocol of the destination server.

The URL, used to distinguish the desired content, always specifies the protocol used by the destination server regardless of the protocol used by the browser to connect to the gateway. In other words, the URL refers only to the destination server's protocol and has no bearing on what protocols may be used in intervening connections.

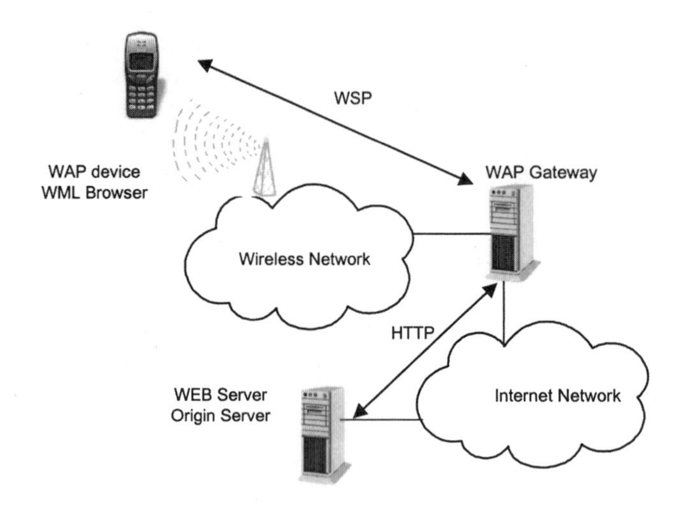

Figure 14.3: WAP network model

In addition to performing protocol conversion by translating requests from WSP into other protocols and the responses back into WSP, the gateway also performs content conversion.

The WAP Gateway utilises Internet standards such as XML, User Datagram Protocol (UDP), and IP. Many of the protocols are based on Internet standards such as HTTP and TLS but have been optimised for the unique constraints of the wireless environment: low bandwidth, high latency and less connection stability.

Internet standards such as HTML, HTTP, TLS and Transmission Control Protocol (TCP) are inefficient over mobile networks, requiring large amounts of mainly text-based data to be sent. Standard HTML content cannot be effectively displayed on the small-size screens of pocket-sized mobile phones and pagers.

The WAP Gateway utilises binary transmission for greater compression of data and is optimised for long latency and low bandwidth. WAP sessions cope with intermittent coverage and can operate over a wide variety of wireless transports.

WML and WMLScript are used to produce WAP content. They make optimum use of small displays, and navigation may be performed with one hand. WAP content is scalable from a two-line text display on a basic device to a full graphic screen on the latest smart phones and communicators.

The lightweight WAP protocol stack is designed to minimise the required bandwidth and maximise the number of wireless network types that can deliver WAP content. These include GSM 900, 1,800 and 1,900 MHz; Interim Standard (IS)-136; Digital European Cordless communication (DECT); Time-Division Multiple Access (TDMA), Personal Communications Service (PCS), and Code Division Multiple Access (CDMA). All network technologies and bearers will be supported, including SMS, USSD, Circuit-Switched Cellular Data (CSD), Cellular Digital Packet Data (CDPD), and GPRS[4].

14.2.2 The WAP Stack

Figure 14.4 presents the Internet and WAP stack.

The wireless session protocol
The WSP layer provides a lightweight session layer to allow efficient exchange of data between applications.

WSP provides the application layer of WAP with a consistent interface for two session services.

1 The first is a connection-oriented service that operates above the transaction layer protocol WTP.
2 The second is a connectionless service that operates above a secure or non-secure datagram service Wireless Datagram Protocol (WDP).

The WSPs currently consist of services suited for browsing applications (WSP/B). They provide the following functions:

- HTTP/1.1 functionality and semantics in a compact over-the-air encoding.
- Long-lived session state.
- Session suspend and resume with session migration.
- A common facility for reliable and unreliable data push.

[4] General Packet Radio Service: Hardware upgrade for TDMA networks that gives each subscriber up to eight 14.4 kbit/s channels and employs packet switching to use bandwidth more efficiently.

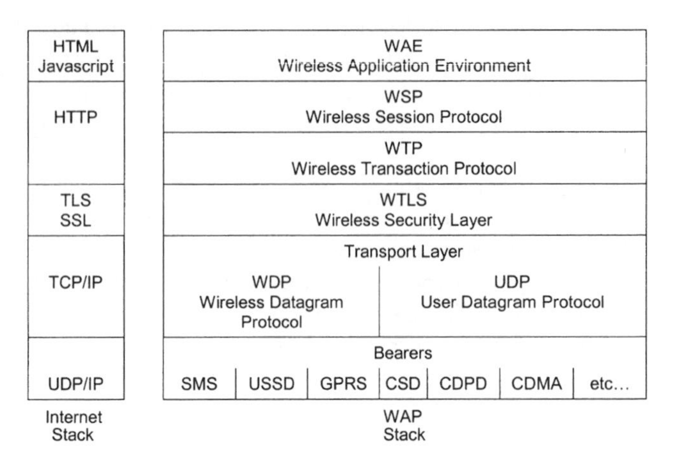

Figure 14.4: Internet and WAP stack

- Protocol feature negotiation.

The WSP protocols are optimised for low-bandwidth bearer networks with relatively long latency. WSP/B is designed to allow a WAP proxy to connect a WSP/B client to a standard HTTP server.

The wireless transaction protocol
The WAP Transaction Protocol (WTP) layer provides transaction support, adding reliability to the datagram service provided by WDP.

The WTP runs on top of a datagram service and provides a light-weight transaction-oriented protocol that is suitable for implementation in 'thin' clients (mobile stations). WTP operates efficiently over secure or non-secure wireless datagram networks and provides the following features:

- Three classes of transaction service:

  1 Unreliable one-way requests.
  2 Reliable one-way requests.
  3 Reliable two-way request-reply transactions.

- Optional user-to-user reliability – the WTP user triggers the confirmation of each received message.

- Optional out-of-band data on acknowledgements.

- PDU concatenation and delayed acknowledgement to reduce the number of messages sent.

- Asynchronous transactions.

The wireless transport layer security

Wireless Transport Layer Security (WTLS), an optional security layer, has encryption facilities that provide the secure transport service required by many applications, such as e-commerce.

WTLS is a security protocol based upon the industry-standard Transport Layer Security (TLS) protocol, formerly known as Secure Sockets Layer (SSL). WTLS is intended for use with the WAP transport protocols and has been optimised for use over narrow-band communication channels. WTLS provides the following features:

- Data integrity: WTLS contains facilities to ensure that data sent between the terminal and an application server is unchanged and uncorrupted.

- Privacy: WTLS contains facilities to ensures that data transmitted between the terminal and an application server is private and cannot be understood by any intermediate parties that may have intercepted the data stream.

- Authentication: WTLS contains facilities to establish the authenticity of the terminal and application server.

- Denial-of-service protection: WTLS contains facilities for detecting and rejecting data that is replayed or not successfully verified. WTLS makes many typical denial-of-service attacks harder to accomplish and protects the upper protocol layers.

WTLS may also be used for secure communication between terminals, e.g. for authentication of electronic business card exchange.

Applications are able to enable or disable WTLS features selectively depending on their security requirements and the characteristics of the underlying network.

The wireless datagram protocol

The transport layer protocol in the WAP architecture is referred to as the WDP. The transport layer is able to send and to receive messages via any available bearer network, including SMS, USSD, CSD, CDPD, IS–136 packet data, and GPRS. The WDP layer operates above the data capable bearer services supported by the various network types. As a general transport service, WDP offers a consistent service to the upper layer protocols of WAP and communicates transparently over one of the available bearer services.

14.2.3 The Client Stack

Figure 14.5 shows the WAE, which has two logical layers:

1 User agents, which include such items as browsers, phonebooks, message editors, etc; and
2 Services and Formats, which include common elements and formats accessible to user agents such as WML, WMLScript, image formats, vCard and vCalendar formats, etc.

WAE separates services from user agents and assumes an environment with multiple user agents but does not imply or suggest an implementation.

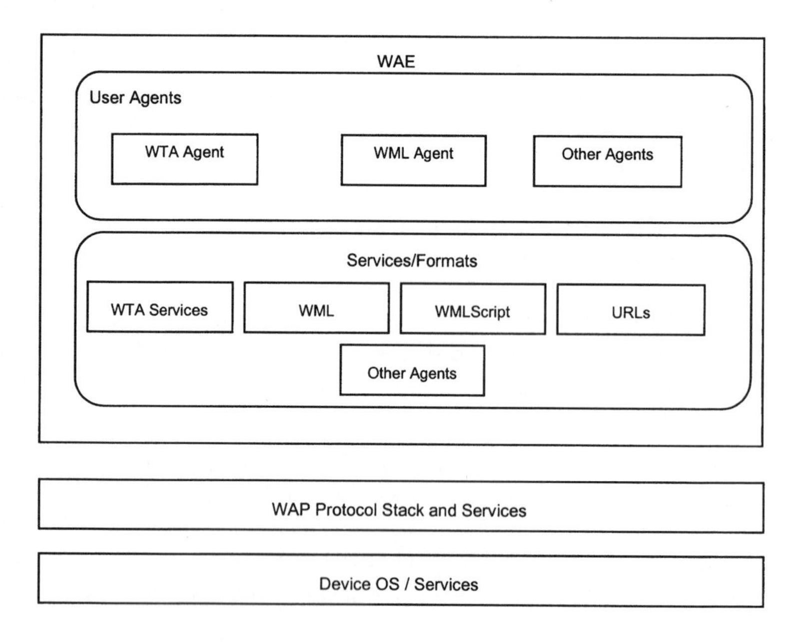

Figure 14.5: WAE Client Components

14.2.4 The WAE User Agents

The WML user agent is a fundamental user agent of the WAE. WAE allows the integration of domain-specific user agents with varying architectures and environments. In particular, a WTA user agent has been specified as an extension to the WAE specification for the mobile telephony environments. The WTA extensions allow authors to access and interact with mobile telephone features (e.g. call control) as well as other applications assumed on the telephones, such as phonebooks and calendar.

WAE does not formally specify any user agent. Features and capabilities of a user agent are left to the implementers. WAE defines only fundamental services and formats that are needed to ensure interoperability among implementations.

14.2.5 The WAE Services and Formats

WML

WML is a tag-based document language. It is an application of a generalised mark-up language. WML shares a heritage with the WWW's HTML and Handheld Device Markup Language (HDML). It is optimised for specifying presentation and user interaction on limited capability devices such as telephones and other wireless mobile terminals.

WML is based on a subset of HDML version 2.0. WML changes some elements adopted from HDML and introduces new elements, some of which have been modelled on similar elements in HTML. The resulting WML implements a card and deck metaphor. It contains constructs allowing the application to specify documents made up of multiple *cards*. An interaction with the user is described in a set of cards, which can be grouped together into a document (commonly referred to as a *deck*). Logically, a user navigates through a set of WML cards. The user navigates to a card, reviews its contents, may enter requested information, may make choices and then move on to another card. Instructions embedded within cards may invoke services on origin servers as needed by the particular interaction. Decks are fetched from origin servers as needed. WML decks can be stored in 'static' files on an origin server, or they can be dynamically generated by a content generator running on an origin server. Each card, in a deck, contains a specification for a particular user interaction. WML allows presentation on a wide variety of devices and vendors can use their own MMIs.

WML has the following features:

- Support for Text and Images.
- Support for User Input.
- Navigation and History Stack.
- International Support (The document character set for WML is the Universal Character Set of ISO/IEC-10646).
- MMI Independence.

- Narrow-band Optimisation (This includes the ability to specify multiple user interactions in one network transfer. It also includes a variety of state management facilities that minimise the need for origin server requests. It provides out-of-band mechanisms for client-side variable passing without having to alter URLs).

- State and Context Management (The state of the variables can be used to modify the contents of a parameterised card without having to communicate with the server).

WMLScript

WMLScript is a lightweight procedural scripting language. WMLScript is loosely based on a subset of JavaScript™

WMLScript provides the application programmer with a variety of interesting capabilities:

- The ability to check the validity of user input before it is sent to the content server.

- The ability to access device facilities and peripherals.

- The ability to interact with the user without introducing round-trips to the origin server (e.g. display an error message).

Key WMLScript features include:

- JavaScript™-based scripting language (for a developer it is very easy to learn and to use WMLScript).

- Procedural Logic (WMLScript adds the power of procedural logic to WAE).

- Event-based.

- Compiled implementation.

- Integrated into WAE.

- International Support.

One objective in designing the WMLScript language was to be close to core JavaScript™. In particular, WMLScript was based on the ECMA-262 Standard 'ECMAScript Language Specification'. The originating technologies for the ECMA Standard include many technologies most notably JavaScript™ and JScript™. WMLScript is not fully compliant with ECMAScript. The standard has been used only as the basis for defining WMLScript language.

URL

WAE assumes a rich set of URL services that user agents can use. In particular, WAE relies heavily on HTTP and HTML URL semantics. In some cases, WAE components extend the URL semantics, such as in WML, where URL fragments have been extended to allow linking to particular WMLScript functions.

WAE content formats

WAE includes a set of agreed content formats that facilitate interoperable data exchange. The method of exchange depends on the data and the targeted WAE user agents. The two most important formats defined in WAE are encoded WML and WMLScript bytecode formats.

WAE defines WML and WMLScript encoding formats that make transmission of WML and WMLScript more efficient as well as minimising the computational efforts required by the client.

14.2.6 The Gateway

A WAP Gateway includes two functions:

1 Protocol Gateway: the protocol gateway translates requests from the WAP protocol stack to the WWW protocol stack (HTTP and TCP/IP).
2 Content Encoders and Decoders: the content encoders translate Web content into compact encoded formats to reduce the size and number of packets travelling over the wireless data network.

This infrastructure ensures that mobile terminal users can browse a variety of WAP content and applications regardless of the wireless network they use. Application authors are able to build content services and applications that are network and terminal independent, allowing their applications to reach the largest possible audience. Because of the WAP proxy design, content and applications are hosted on standard WWW servers and can be developed using proven Web technologies such as CGI scripting.

The WAP Gateway decreases the response time to the handheld device by aggregating data from different servers on the Web, and caching frequently used information. The WAP Gateway can also interface with subscriber databases and use information from the wireless network, such as location information, to dynamically customise WML pages for a certain group of users.

Figure 14.6 shows a gateway stack.

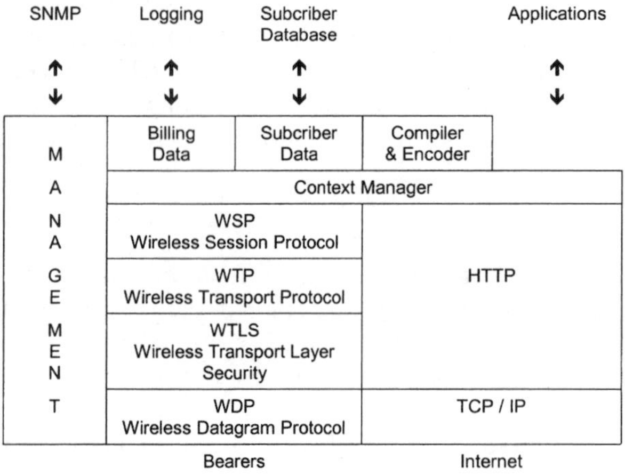

SNMP Logging Subcriber Database Applications

MANAGEMENT

Billing Data | Subcriber Data | Compiler & Encoder

Context Manager

WSP Wireless Session Protocol

WTP Wireless Transport Protocol

WTLS Wireless Transport Layer Security

HTTP

WDP Wireless Datagram Protocol

TCP / IP

Bearers Internet

Figure 14.6: Gateway design

14.2.7 Internationalisation

The WAE architecture is designed to support mobile terminals and network applications using a variety of languages and character sets. This work is collectively described as internationalisation (referred to as I18N). It is a design goal of WAE to be fully global in its nature in that it supports any language.

The WAE architecture makes the following assumptions regarding I18N:

- WAE user agents will have a current language and will accept content in a set of well-known character encoding sets.

- Origin server-side applications can emit content in one or more encoding sets and can accept input from the user agent in one or more encoding sets.

14.2.8 Security

WAE dominates WTLS where services require authenticated and/or secure exchanges. In addition, both WML and WMLScript include access control constructs that communicate URL-based access restrictions to the client.

WAE also supports HTTP 1.1 basic authentication where a server can authenticate the client based on a username and password[5].

[5] Entrust Technologies (Plano, TX), PKI firm, has signed a non-exclusive deal with Nokia which will allow users to make secure online purchases via their phones using WAP.

14.3 The Wireless Telephony Application (WTA)

14.3.1 WTA and WAE Users Agents

The WTA user-agent is an extension to the standard WML user-agent with the addition of capabilities for interfacing with mobile network services available to a mobile telephone, e.g. setting up and receiving phone calls. Figure 14.7 describes one possible configuration of the WTA framework. However, this specification solely defines the components contained in the client.

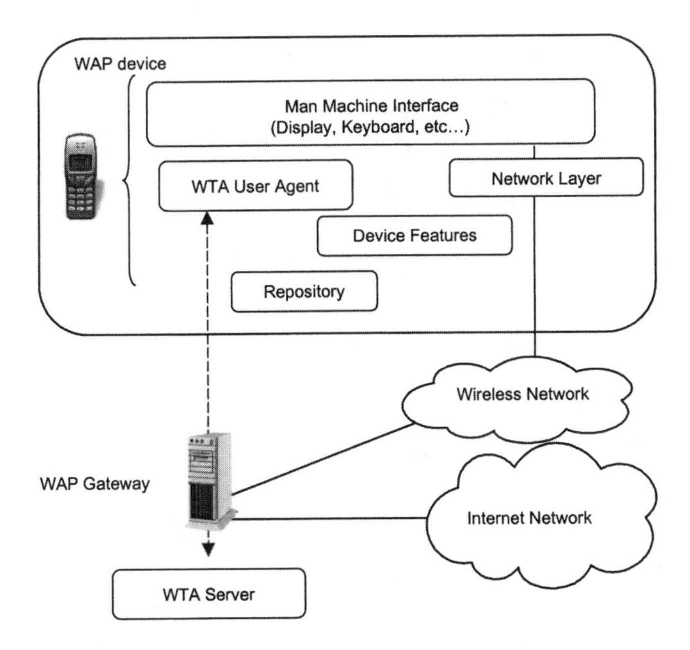

Figure 14.7: WTA design

In order to support telephony functions for a WAE user-agent, a special WTA Library, the WTAI Public Library has been defined. This library contains functions which can be called from any WAE application (Figure 14.8) and provides access to telephony functionality as an aide to the user experience. For instance it allows WML authors to include 'click to phone' functionality within their content, to save users from typing a number using the default Man Machine Interface.

The repository (persistent storage) and the Telephony Application Interface (WTAI) interact with each other and other entities in a WTA-capable mobile client device. The WTA user-agent is able to retrieve content from the repo-

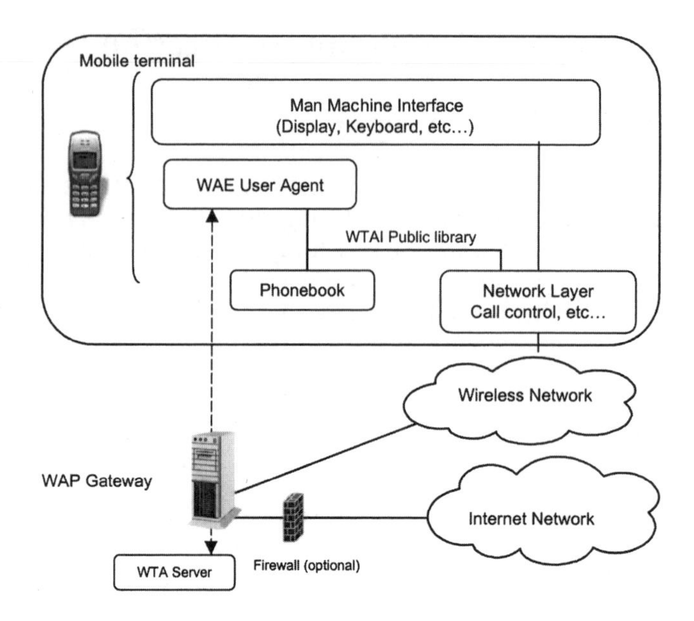

Figure 14.8: WAE user and the WTA public library

sitory and WTAI ensures that the WTA user-agent can interact with mobile network functions (e.g. setting up calls) and device specific features (e.g. manipulating the phonebook). The WTA user-agent receives 'network events' that can be bound to content, thus enabling dynamic telephony applications.

The network events that will be available to the WTA user-agent are those that are the result of actions taken by services running in the WTA user-agent itself. Telephony events initiated from outside the device are also passed to the WTA user-agent, as are network text message events that are not explicitly directed towards another user-agent (e.g. events intended for a SIM). This means, for instance, that network events caused by the WML user-agent will not affect the WTA user-agent.

Figure 14.8 illustrates how the WAE user-agent and WTAI Public Library interact with each other and other entities in a WTA-capable mobile client. The WAE user-agent retrieves its content only via the WAP gateway and only has access to the WTAI Public Library functions. These functions support simple functionality such as the ability to place a call, but do not allow fully featured telephony control. Only a WTA user-agent is able to fully control the telephony features of the device.

The WTA server can be thought of as a web server delivering content requested by a client. Like an Internet web browser, a WTA user-agent uses URLs to reference content on the WTA server.

By referencing applications on a WTA server it is possible to create services that use URLs to interact with the mobile network and other entities (e.g. a voice mail system). Thus, the concept of referencing applications on a WTA server provides a simple but powerful model for seamless integration of services in, e.g. the mobile network with services executing locally in the WAP client.

The repository is a continuous storage module within the mobile terminal that may be used to eliminate the need for network access when loading and executing frequently used WTA services. The repository also addresses the issue of ensuring that time-critical WTA events are handled in a timely manner by a WTA service developer.

14.3.2 WTA Services

WTA services are what the end-user ultimately experiences from using the WTA framework. A WTA service appears to the client in the form of various content formats, e.g. WML, WMLScript, etc. The WTA user-agent executes content that is continuously stored in the client's repository or content retrieved from a WTA server. The framework also allows the WTA user-agent to act on events from the mobile network (e.g. an incoming call).

The WTA user-agent essentially executes content within the boundary of a well-known context. The term service is used to define the extent of a context and its associated content. Initiation of a new context is a defining start of a service. Termination of a context marks the defining end of a service. Figure 14.9 shows the possible ways of initiating a WTA service in the WTA user-agent.

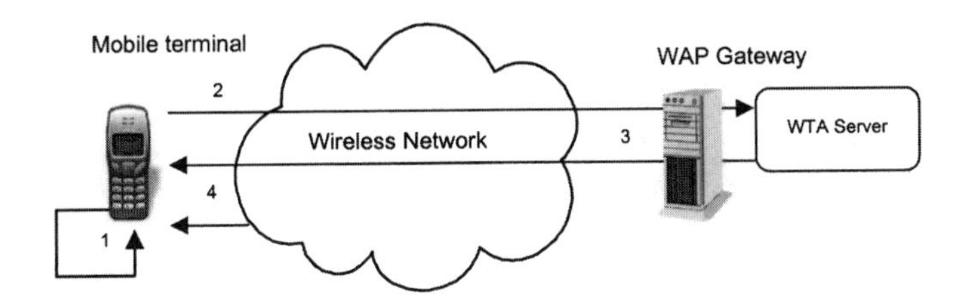

Figure 14.9: WTA service invocation

1 URL selection (via the repository).
2 Access to a URL (via the WTA server).

3 Service Indication (Push).
4 Network event.

WTA services are created using WML and WMLScript. From a WMLScript, telephony functions can be accessed through the Wireless Telephony Applications Interface (WTAI). WTAI also provides access to telephony functions from WML by using URIs [RFC1630]. URIs form a unifying naming model to identify features independently of the internal structure of the device and the mobile network. The WTA services reside on the WTA server. The client addresses WTA services by using URLs [RFC1738]. Examples of WTA services include:

- Extended set of user options for handling incoming calls (incoming call selection):
- The service is started when an incoming call is detected by the client. A menu of user options is presented to the user. Examples of options could be:
    - Accept call.
    - Redirect to voice mail.
    - Redirect to another subscriber.
    - Send special message to caller.
- Voice mail:
- The user is notified that he has new voice mails, and retrieves a list of them from the server. The list is presented on the client's display. When a certain voice mail has been selected, the server sets up a call to the client and the user listens to the selected voice mail.
- Call subscriber from message list or log:
- When a list of voice, fax or e-mails or any kind of call log is displayed the user has the option of calling the originator of a selected entry in the list or log.

14.3.3 WTA Security

The security between the client and the WTLS connection end point is ensured using WTLS class 2 with the use of server certificate authentication. In the WTLS security model, WTLS certificates are stored in the WAP client, the WTLS session based on these certificates is also used for WTA.

The WTLS security model extends from client to the gateway (Figure 14.10).

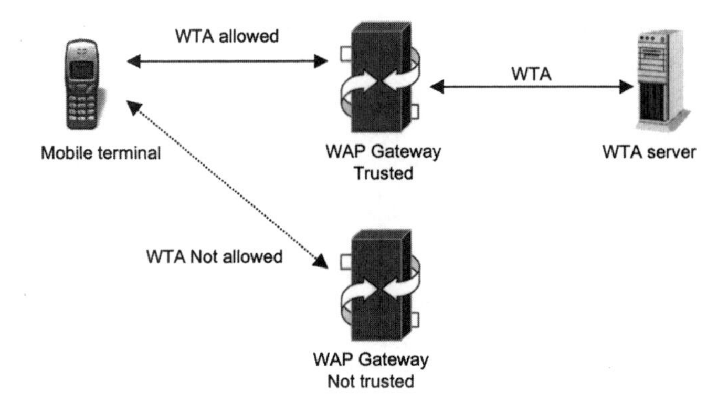

Figure 14.10: WTA security model

The service provider (or an entity delegated by the service provider) shall supervise the gateway in order to ensure that a 'secure' trust relationship is maintained between the gateway and the WTA server. The WTLS security model uses certificates stored in the WAP client to ensure that only trusted gateways can be used in order to access telephony resources in the device. Using a transport layer security model, any WTA service can be delivered by the service provider (or an entity delegated by the service provider) through a secure pipe from the gateway to the client. WAP gateway to WTA server security is delegated to the implementation. For connections over the Internet, SSL/TLS may be used.

The client must only allow connections for WTA services to trusted gateways. This is achieved by the identity, authenticated by its certificate, of the specified gateways being an identity specifically allowed for such connections. The WTLS layer (WTLS Session) secures the connection between the client and the WAP gateway, the content (identified by WTLS authentication) can be executed directly by the WTA user-agent. There is no need for any additional processing of the content in order to determine its validity and all operations can be trusted.

14.4 The Future of WAP

The tremendous surge of interest and development in the area of wireless data in recent times has caused worldwide operators, infrastructure and terminal manufacturers, and content developers to collaborate on an unprecedented scale, in an area notorious for the diversity of standards and protocols. The collaborative efforts of the WAP Forum have devised, and continue to develop, a set of protocols that provide a common environment for the development of advanced telephony services and Internet access for the wireless market. If the WAP protocols were to be as successful as the Transmission Control Protocol (TCP)/Internet Protocol (IP), the boom in mobile communications would be phenomenal. Indeed, the WAP browser should do for mobile Internet what Netscape did for the Internet.

Industry players from content developers to operators can explore the vast opportunity that WAP presents. As a fixed-line technology, the Internet has proved highly successful in reaching the homes of millions worldwide. However, mobile users until now have been forced to accept relatively basic levels of functionality, over and above voice communication and are beginning to require the industry to move from a fixed to a mobile environment, carrying the functionality of a fixed environment with it.

Initially, services are expected to run over the well-established SMS bearer, which will dictate the nature and speed of early applications. Indeed, GSM currently does not offer the data rates that would allow mobile multimedia and Web browsing. With the advent of GPRS, which aimed at increasing the data rate to 115 kbps, as well as other emerging high-bandwidth bearers, the reality of access speeds equivalent to or higher than that of a fixed-line scenario become evermore believable. GPRS is seen by many as the perfect partner for WAP, with its distinct time slots serving to manage data packets in a way that prevents users from being penalised for holding standard circuit-switched connections.

14.4.1 Handset Manufacturers and WAP Services

It is expected that mobile terminal manufacturers will experience significant change as a result of WAP technology – a chance that will have an impact on the look and feel of the hardware they produce. The main issues faced by this arm of the industry concern the size of mobile phones, power supplies, display size, usability, processing power and the role of Personal Digital Assistants (PDAs) and other mobile terminals.

With over 75 percent of the world's key handset manufacturers already involved in the WAP Forum and announcing the impending release of WAP-compatible handsets, the drive toward new and innovative devices is quickly gathering pace. The handsets themselves will contain a microbrowser that will serve to interpret the byte code (generated from the WML/WMLS content) and display interactive content to the user.

The services available to users will be wide-ranging in nature, as a result of the open specifications of WAP, their similarity to the established and accepted Internet model, and the simplicity of the WML/WMLS languages with which the applications will be written. Information will be available in push and pull functionality, with the ability for users to interact with services via both voice and data interfaces. Web browsing as experienced by the desktop user, however, is not expected to be the main driver behind WAP as a result of time and processing constraints.

Real-time applications and services demand small key pieces of information that will fuel the success of WAP in the mobile marketplace. Stock prices, news, weather, and travel are only some of the areas in which WAP will provide services for mobile users. Essentially, the WAP application strategy involves taking existing services that are common within a fixed-line environment and tailoring them to be purposeful and user-friendly in a wireless environment.

Empowering the user with the ability to access a wealth of information and services from a mobile device will create a new battleground. Mobile industry players will fight to provide their customers with sophisticated, value-added services. As mobile commerce becomes a more secure and trusted channel by which consumers may conduct their financial affairs, the market for WAP will become even more lucrative.

14.4.2 WAP in the Competitive Environment

Competition for WAP protocols could come from a number of sources:

Subscriber Identity Module (SIM) toolkit-The use of SIMs or smart cards in wireless devices is already widespread and used in some of the service sectors. Windows CE-This is a multitasking, multithreaded operating system from Microsoft designed for including or embedding in mobile and other space-constrained devices. JavaPhone™-Sun Microsystems is developing PersonalJava™ and a JavaPhone™ API, which is embedded in a Java™ virtual machine on the handset. NEPs will be able to build cellular phones that can download extra features and functions over the Internet; thus, customers will no longer be required to buy a new phone to take advantage of improved features.

The advantages that WAP can offer over these other methods are the following:

- open standard;
- vendor independent and network-standard independent;

- transport mechanism–optimised for wireless data bearers;
- application downloaded from the server, enabling fast service creation and introduction, as opposed to embedded software.

14.5 References

RFC2068 "Hypertext Transfer Protocol – HTTP/1.1", R. Fielding et al., January 1997.

RFC2396 "Uniform Resource Identifier (URI): Generic Syntax", T. Berners-Lee et al., August 1998.

VCARD vCard – The Electronic Business Card; version 2.1; The Internet Mail Consortium (IMC), September 18, 1996.

VCAL vCalendar – the Electronic Calendaring and Scheduling Format; version 1.0; The Internet Mail Consortium (IMC), September 18, 1996.

WAE "Wireless Application Environment Specification", WAP Forum, November 4, 1999. URL: http://www.wapforum.com/

WAP "Wireless Application Protocol Architecture Specification", Wireless Application WAP Forum, April 30, 1998. URL: http://www.wapforum.com/

WBXML "Binary XML Content Format Specification", WAP Forum, November 4, 1999. URL: http://www.wapforum.com/

WML "Wireless Markup Language Specification", WAP Forum, November 4, 1999. URL: http://www.wapforum.com/

WSP "Wireless Session Protocol", WAP Forum, November 5, 1999. URL: http://www.wapforum.com/

WTA "Wireless Telephony Application", WAP Forum, November 8, 1999. URL: http://www.wapforum.com/

WTLS "Wireless Transport Layer Security", WAP Forum, November 5, 1999. URL: http://www.wapforum.com/

WTP "Wireless Transaction Protocol", WAP Forum, June 11, 1999. URL: http://www.wapforum.com/

XML "Extensible Markup Language (XML), W3C Proposed Recommendation, December 8, 1997, PR-xml-971208", T. Bray et al., December 8, 1997.

Enhanced Data Rates for GSM Evolution

15

15.1 Introduction

This chapter considers evolutionary changes in GSM, to enable it to provide a variety of Third Generation (3G) concepts. These evolutionary changes are presented in the form of options that would:

- provide a high quality voice service for indoor and pedestrian systems such as cellular office systems and personal base stations;
- support enhanced bit-rate wireless packet data access to the Internet as well as circuit data access;
- provide smart antenna technology to improve coverage, quality and capacity;
- automatically assign operating frequencies and provide for dynamic channel reconfiguration;
- support microcellular arrangements to provide a low cost and high capacity service in dense areas;
- support a future high-speed packet data access mode through a wide-band system.

15.2 Enhanced Voice Services

There is increasing interest in combining GSM or DCS-1800 technology with DECT technology in a single multimode handset. Standards are being developed to support network interoperation between GSM based cellular systems and privately owned DECT based keysets or PBXs. Ericsson and others are developing dual-mode GSM/DECT handsets. A number of service operators are planning to deploy GSM based PCS technology in the USA, and they are likely to make dual-mode GSM/DECT based technologies available in the USA. In Asia, there is interest in dual-mode GSM/PHS or PDC/PHS handsets. This approach potentially supports wireline quality speech with low transmission delay in private indoor environments while permitting a single personal terminal to support wireless access for both indoor private wireless access and widespread cellular access. A disadvantage of this approach is the need for a handset to support two air interfaces. This approach uses separate spectrum for the public and private systems.

The 3G standards offer the opportunity for integrating macrocellular and indoor wireless access in a seamless fashion while using a low-cost handset. The 3G handsets will include private wireless access capabilities that are used in residential and business environments and their digital speech coders will be common.

The requirement for high quality speech is likely to be most pronounced for the microcellular, residential cordless, wireless keyset and wireless PBX environments where speech quality will be compared directly with wireline access.

One approach to improving voice quality is to improve the performance of voice coders at 8 kbps. The Algebraic Code Excited Linear Predictive (ACELP) speech coding standard provides significant improvement over the initial 8 kbps Vector Excited Linear Predictor (VSELP) (IS-54 is the TIA standard for North American digital cellular communications) speech coder used with TDMA, but significant transmission delay remains.

A second approach to improving voice quality is to allocate multiple time-slots to individual users in order to support voice coders at 16 kbps (two time-slots out of three for one frequency channel) or 24 kbps (all three time-slots for a channel). Unfortunately this has a substantial impact on capacity, it limits the ability of an economical terminal to perform the required signal strength measurements for mobile-assisted-handover, and it requires terminals to include a duplexer.

A third approach to improving the voice quality is to introduce 8-PSK (Quartenary Phase Keying) or 16-QAM (Quadrature Amplitude Modulation) modulation with efficient channel coding and diversity techniques to provide sufficient robustness in the presence of transmission impairments. This approach could permit the introduction of a higher rate voice coder.

15.2.1 Speech Coding

Two higher rate speech coders are considered. ITU Recommendation G.728 and the GSM Enhanced Full Rate coder both give improved speech quality compared to an IS-641 coder. In this section both alternatives are considered and compared with IS-641 on the basis of complexity, voice quality and delay. The G.728 Low Delay Code Excited Linear Prediction (LD-CELP)voice coder at 16 kbps provides the possibility of an advanced mode of operation with toll voice quality and relatively low delay.

The GSM Enhanced Full Rate speech coder at 12.2 kbps was designed as a replacement for the original GSM speech coder in PCS-type systems. It is slightly more complex than the IS-641 speech coder, owing to a larger excitation search.

Wireless phones are designed so that the microphone is further from the talker's mouth than for fixed phones. They also tend to be used in higher ambient noise backgrounds than conventional wired phones that are used in a home or office. As a result, noisy input speech is more of a problem for wireless phones and the performance of a digital speech coder in these conditions is significant for its acceptance in the marketplace. In addition to speech with music in the background, music-on-hold is an important quality characteristic for these coders.

Another important factor in voice quality is performance on noisy channels. G.728 was designed to withstand bit errors up to a rate of 1% random bit errors. It was not designed to be combined with a channel coder. Consequently, every bit needed to be equally insensitive to bit errors. However, its performance at 1% bit error rate is poor. Wireless channels often have error rates higher than this. As a result, G.728 is at a disadvantage compared with IS-641 and GSM-EFR. These coders were designed to be used with a channel coder. Consequently, while some of their bits are highly sensitive to bit errors, they are well protected by the channel code. If any of the most sensitive bits are in error, a frame erasure is declared. Having unequal error sensitivity is advantageous when only a small number of bits is available for error protection. The disadvantage of G.728 is further exacerbated by its higher bit rate. This means fewer bits are available for error protection. Delay is also an important parameter for which the G.728 vocoder is the best.

15.2.2 Modulation Channel Coding

In order to support the higher data rate requirement of the 12 and 16 kbps speech coders and to meet the requirement of three users in a 30 kHz channel, it is necessary to change the modulation from pi/4-DQPSK to a higher modulation level such as 8-PSK. By changing only those symbols which would carry user data from pi/4-DQPSK to a higher modulation level, the higher layer protocols can be left largely unchanged. This would permit a high degree of commonality between various modes.

The maximum possible signalling rate is 58.5 kbps with 8-PSK using the IS-136 format. This assumes that the same number of symbols is devoted to signalling, and to other overheads, such as channel estimation, as in the existing standard. The 58.5 kbps data rate with 8-PSK is enough to support up to 16 kbps for three voice users. Data users can be supported at 40–50 kbps with 8-PSK and multi-slot operation and at 50–60 kbps with 16-QAM and multi-slot operation. The remaining data rate could be used for channel coding to provide robustness and for additional channel estimation that is needed for data transmission in fast fading channel environments.

To provide protection against interference, fading and noise, channel coding and/or some form of diversity is necessary. A major benefit of channel coding is its ability to provide diversity in a fast fading environment thus reducing the need for any other form of explicit diversity. This coding can provide some robustness only against interference and thermal noise.

15.3 Enhanced Speed for TDMA Data Services

TDMA provides 9.6 kbps circuit data and FAX access based on the wireless 2G standards. Enhanced data rates will be important in the future, particularly for web browsing on the Internet or on corporate Intranets.

Enhanced access speed will require additions to the existing standards. The first step is likely to be multi-slot operation which can support rates up to 28.8 kbps.

A second step is to introduce over-the-air packet access for TDMA in addition to circuit mode access.

A further step will be to introduce 8-PSK and/or 16-QAM modes to support data rates up to about 57.6 kbps. Even higher bit rates may be possible based on transmit and receive diversity. Wireless data access to the Internet is often likely to come from stationary but widespread users with laptop computers or PDAs. Under those conditions, simple pre-selection diversity, as proposed for high quality pedestrian voice service, could significantly improve downlink performance since fading rates will be slow. The introduction of multi-slot operation brings several challenges:

- increased signal processing requirements in terminals;
- the possible requirement for a duplexer to support simultaneous transmit and receive for terminals;
- a requirement to continue to support Mobile-Assisted-Handovers;
- flexibility in the network to process both single-slot and multi-slot access.

Since data-only operation removes the need to perform speech coding, which is a major signal processing demand for terminals, multi-slot data operation should not increase signal processing requirements significantly.

Duplexer requirements are a concern for 1.9 GHz PCS operation, because the upband and the downband are each 60 MHz wide with only a 20 MHz band separating them, so low-loss duplexers are a problem. This may motivate consideration of MAC protocols which avoid requiring a terminal to receive and transmit simultaneously by scheduling periods at the base station for a terminal to transmit.

A full-rate TDMA channel is arranged with one time-slot out of three used for the terminal receiver every 20 ms and with one time-slot used for the transmitter that overlaps the two unused receive time-slots.

To avoid a duplexer requirement for multi-slot operation when the data transmission is primarily downlink, the MAC protocol could reserve two time-slots (13.3 ms) every 100 ms for a terminal to acknowledge incoming data on the forward link. This would also provide time for terminals to make Mobile-Assisted-Handover measurements. The cost would be a reduction in throughput by about 13.3% and an increase in latency due to sparse acknowledgements. Such a scheme also requires a protocol for varying the ratio of uplink and downlink bandwidth robustly in real-time. One way to achieve robustness would be to make the base station the master with a fixed superframe of 100 ms with downlink transmissions beginning at the start of the superframe. A terminal would always be allocated one time-slot at the end of the frame for transmission of acknowledgements, data and control information, but it could request more if required. For multi-slot circuit data operation, packet data access can include protocol provision at the base station to schedule time-slots for terminals to perform MAHO and acknowledgements without requiring duplexers at the terminal.

15.4 Smart Antenna Technology

Smart antennas and adaptive antenna arrays have recently been of increasing interest to improve the performance of cellular radio systems. Smart antennas include a large number of techniques that attempt to enhance a desired signal and suppress interfering signals. While adaptive antenna arrays and steerable beam antennas have been used for military applications for decades, only in the last few years have these techniques begun to be applied to commercial systems.

Research on adaptive antenna arrays for cellular systems dates from the early to mid 1980s but R&D on smart and adaptive antennas for cellular systems has intensified only in the last few years. In 1995, Nortel introduced smart antenna technology for PCS-1900 systems. Other companies such as Metawave have introduced similar technology, and the European Advanced Communications Technologies and Services (European Community Projects) (ACTS) TSUNAMI (a European Community project Technology in Smart Antennas for Universal Advanced Mobile Infrastructure) project is considering adaptive antennas for third generation wireless systems.

Smart or adaptive antennas are receiving much attention because of several critical factors. Cellular networks are becoming increasingly crowded and smart antennas offer the possibility of improving quality while increasing capacity by suppressing interference and supporting operation with lower cellular reuse factors. The new PCS bands at 1.9 GHz are subject to increased path loss that approaches 10 dB relative to the cellular bands. This means that the new PCS systems will be under significant pressure to provide adequate coverage, and techniques to extend range will be very important. Since wireless access has become dominated by handsets with limited transmit power, the uplink is generally the factor which limits range. Finally, the key technologies required for smart antennas, digital signal processing and efficient RF systems, are rapidly advancing. DSP chips are becoming available with throughputs on the order of 100 MIPS that require only hundreds of milliwatts of power, and this means that smart antennas depending on complex algorithms are becoming commercially feasible. For instance, wideband digitisation and digital receiver techniques are becoming practical for base stations, and high-power multicarrier linear amplifiers are now attractive alternatives to single-carrier amplifier and cavity-combiner approaches. Smart antennas offer the potential to lower the cost of the cellular infrastructure. One estimate suggests that smart antennas can reduce the infrastructure cost for a cellular or PCS system by up to 60%.

There are two basic classes of smart antennas that are being considered for cellular systems:

1 switched beam antennas;
2 adaptive antenna arrays.

Switched beam antennas: a typical switched beam antenna has four 30° beams within a conventional 120° sector for a base station. This antenna consists of four collinear elements placed in a line in front of a reflector to individually form 120° patterns. The collinears are spaced at about one half wavelength apart and are fed through a four-port Butler matrix. This arrangement provides four 30° beams that each have a peak gain that is about 6 dB higher than any collinear individually. By combining such an arrangement with an RF switching matrix and a scanning receiver to choose the beams with the best signals from a mobile, uplink sensitivity can be improved for a desired signal and interference that is outside the selected beams is suppressed. A variation on this approach for cellular and PCS applications is to replace the Butler matrix with processing that allows the peak of a beam to be pointed at any angle within the 120° sector.

Adaptive antenna arrays: In the adaptive antenna array case, four conventional 120° sector antennas may be spaced uniformly with a total aperture of 10–20 wavelengths (5–10 feet at 1.9 GHz). A goal of this antenna configuration is to achieve minimum correlation in fading between the four elements to maximise diversity gain. The probability of simultaneous fades is then low. The signals from the four antenna elements are individually processed and combined using digital signal processing. The four signals are combined using weights that maximise the desired signal and suppress any interfering signals. There are many possible algorithms for generating the combining weights and there are also many different antenna element configurations that could be considered.

15.5 Automatic and Dynamic Channel Assignment

Today's TDMA systems are almost entirely based on conventional Fixed Channel Assignment (FCA). The exceptions to this are emerging applications of wireless office systems, personal base stations and microcellular systems. On the other hand, a large body of work exists which has examined the possible use of Automatic Channel Assignment (ACA) and Dynamic Channel Assignment (DCA) for cellular systems, and DCA has been considered for mobile systems from the early days of research on cellular telephony. The motivations for DCA include spectrum efficiency, dealing with 'hot spots', and reducing manual frequency planning.

DCA schemes have been implemented for indoor, low-tier and pedestrian technologies such as DECT and Personal Handyphone System (PHS). Several factors suggest that DCA will become a core capability for TDMA systems:

1. the emergence of hierarchical cellular systems;
2. the development and deployment of automatically tuned cavity combiners and multicarrier amplifiers;
3. the experience learned from early adopters of DCA techniques.

TDMA uses many channels each of which is only 30 kHz wide. This makes TDMA particularly well suited to dynamic channel allocation for microcells, personal base stations and wireless office systems within a macrocellular system that uses either FCA or DCA. With channels that are only 30 kHz wide, microcells and indoor systems that are uncoordinated or loosely coordinated with a macrocellular system can assign channels in small increments. This is particularly useful for the coexistance of a large number of overlapping or nearby systems that are uncoordinated except for frequency assignment arrangements. If microcells and indoor systems make measurements to avoid channels used on nearby macrocells and to coordinate frequency use among themselves, this will encourage some rules for DCA to minimise channel assignment churn and harmful interference. For example, macrocellular DCA should include some rules or inertia in reassigning channels to avoid churn in microcellular and indoor system frequency assignments. Macrocells typically transmit with much higher power than microcells, so microcells can readily measure the signals of nearby macrocells for ACA, but macrocells can not easily measure the signals of nearby microcells, so macrocells must depend on rules or data bases instead of measurements to avoid interfering with microcells upon making new channel assignments.

Two basic types of DCA schemes have been studied for cellular applications:

1. measurement based schemes that make real-time measurements at the mobile and/or base station to find 'unused' channels;
2. network based schemes that use a set of rules to control interference by maintaining minimum reuse distances or a learning process for finding good channels.

With certain assumptions, measurement based schemes can provide very high performance, particularly if channel selection is based on a combination of both mobile and base station measurements. For example, with one proposal, each base station broadcasts a candidate list of low interference channels as seen by its receivers. A mobile scans the base station's candidate list for the lowest interference channel as seen by its receiver, so the resulting channel selection process provides a low interference channel for both the uplink and downlink. A common assumption for measurement based DCA is that all terminals are relatively stationary.

15.6 Microcellular Evolution

Microcells today share much of their architecture with macrocells. In the future, TDMA microcells will be more highly optimised for small cell operation based on several trends including:

- centralisation of functionality;
- miniaturisation;
- power reduction;
- flexibility of interconnections;
- line powering for picocells;
- automatic channel assignment.

The optimisation of microcells for small cell operation will make TDMA more cost effective and flexible for deployment in dense pedestrian areas and in private indoor environments.

Advances in high-speed A/D, D/A, DSP technology, and the availability of high-speed fibre networks may permit microcells that contain only RF processing, A/D, D/A and fibre interconnect circuitry to support the digitisation of an entire RF band of operation. Possible advantages include microcells that are independent of the evolution of modulation and standards.

15.7 Packet Data Wireless Access Using EDGE

The third generation wireless network proposes to provide a broad range of services. Web browsing and information service access, which has caused the recent explosion in Internet usage, is highly asymmetrical in transmission requirements. Only the transmission path to the subscriber needs to be high speed for many applications. Many other services provided over the Internet can also be provided with low to moderate bit rate uplinks. However, large file transfer is an example of an application which benefits from symmetrical high-speed transmission.

The Universal Wireless Communications Consortium (UWCC) has adopted Enhanced Data rates for GSM Evolution (EDGE) based on enhancing GSM packet data technology with adaptive modulation to support packet data access at peak rates up to about 384 kbps. This system will adapt between Gaussian Minimum Shift Keying (GMSK) and 8-PSK modulation with up to eight different channel coding rates, as shown in Table 15.1, to support packet data communication at high-speeds on low interference/noise channels and at lower speeds on channels with heavy interference/noise. EDGE can be deployed with conventional 4/12 reuse as used for GSM (Figure 15.1). However, the use of 1/3 reuse is also proposed to permit initial deployment

Table 15.1: EDGE Modulation schemes

Scheme	Modulation	Bit rate (kbps)
MCS_8	8 PSK	59.2
MCS_7	8 PSK	44.8
MCS_6	8 PSK	29.6
MCS_5	8 PSK	22.4
MCS_4	GMSK	17.6
MCS_3	GMSK	14.8
MCS_2	GMSK	11.2
MCS_1	GMSK	8.8

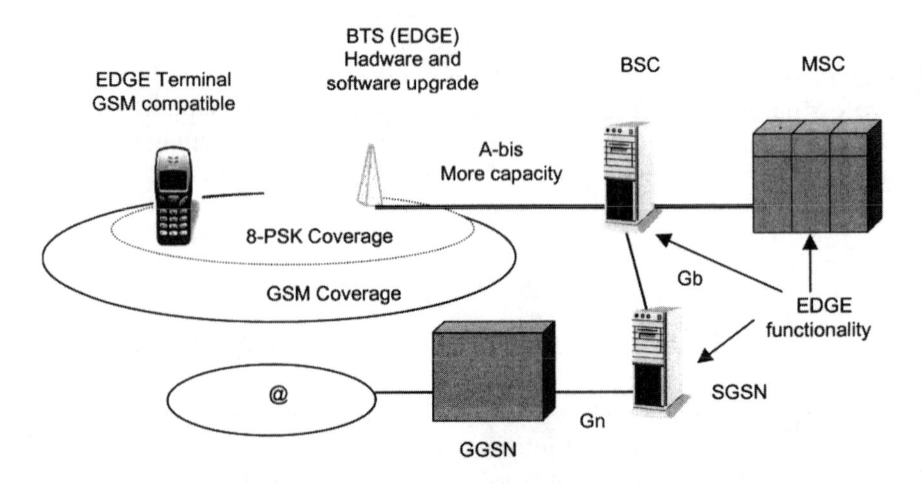

Figure 15.1: EDGE Network elements affected

with only 2 × 1 MHz of spectrum. A combination of adaptive modulation/coding, partial loading and efficient Automatic Repeat Request (ARQ) permits operation with very low reuse factors. Incremental redundancy or hybrid ARQ has been proposed to reduce the sensitivity of adaptive modulation to errors in estimating channel conditions and to improve throughput. One challenge for 1/3 reuse is the operation of control channels which require robust operation. This can be accomplished by synchronising the frame structures of base station transceivers throughout a network and using time-reuse to achieve adequate protection for control channels. GPRS networking is proposed to support EDGE based wireless packet data access.

15.8 Summary

This chapter has considered evolutionary changes to second generation systems to enable them to provide a variety of third generation wireless services. These evolutionary changes include options that would:

- provide high quality voice service for indoor and pedestrian systems such as cellular office systems and personal base stations;
- support enhanced bit-rate packet wireless data access to the Internet as well as circuit data access;
- provide smart antenna technology to improve coverage, quality and capacity;
- automatically assign operating frequencies and provide for dynamic channel reconfiguration;
- support microcellular arrangements to provide low cost and high capacity services in dense areas;
- support future high-speed packet data access modes through a wideband system.

The introduction of a complementary high-speed packet data service will require very high spectrum efficiencies to permit the economical re-allocation of spectrum from existing narrowband services to high-speed operation.

Bibliography

[1] ART, *Perspectives d'évolution à moyen terme du marché français du radio-téléphone*. Etude menée par l'IDATE pour le compte de l'Autorité de regulation des télécommunications, April 1998.

[2] ART, Réponse de l'Autorité de regulation de télécommunications au Livre vert de la Commission européenne sur la convergence des secteurs des télécommunications, des médias et des technologies de l'information, et les implications pour la réglementation, May 1998.

[3] Ballard M., Issenmann E., Moya-Sanchez M., *Intelligent Network Application to Mobile Radio Systems*. Colloque International de Commutation ISS'90, Stockholm, 1990.

[4] Balston D.M., Macario R.C.V., *The Pan-European System: GSM*. Editions Cellular Radio Systems, Artech House, Boston, MA, 1993.

[5] Balston D.M., Cheeseman D., *The Pan-European Cellular Technology*. Editions Personal and Mobile Radio Systems, Peter Peregrinus, London, 1991.

[6] Bezler M. et al., *GSM Base Station System*. Electrical Communication, 1993.

[7] Boehm M., *Chances and Risks for the Pan-European Cellular Mobile Radio Telephone*. FITCE, no. 1, 1988.

[8] Buckingham S., *An Introduction to SMS*. Mobile Lifestreams Ltd, June 1999.

[9] Burst T.P, Ho K.K.Y., Kunzinger F.F., Roberts L.N., Shanks W.L., Tantillo L.A., *Digital Radio for Mobile Applications*. AT&T Technical Journal, July/August 1993.

[10] Calhom G., *Radio Cellulaire Numérique*. Lavoisier Tech&Doc.

[11] CCITT, Livre rouge VI, fascicule VI.7.

[12] CCITT, Projet de recommandation M.3010, *principes d'un réseau de gestion des télécommunications*.

[13] Chang J.J.C., Miska R.A., Shober A.R., *Wireless Systems and Technologies: An Overview*. AT&T Technical Journal, July/August 1993.

[14] Déchaux C., Scheller R., *What are GSM and DCS*. Electrical Communication, 1993.

[15] Duplessis P., Maillard F., *Le système cellulaire numérique paneuropéen de radiotéléphonie avec les mobiles*, Commutation & Transmission, no. 2, 1988.

[16] ETSI, TR 101 375 V1.1.1, Security Algorithms Group of Experts (SAGE), Report on the specification, evaluation and usage of the GSM GPRS Encryption Algorithm (GEA), September 1998.

[17] ETSI, TS 100 629 V6.1.0, Digital Cellular Telecommunications System (Phase 2+), Subscriber Data Management, August 1998.

[18] ETSI, TS 101 348 V6.2.0, Digital Cellular Telecommunications System (Phase 2+), General Packet Radio Service (GPRS), Interworking between the Public Land Mobile Network (PLMN) Supporting GPRS and Packet Data Networks (PDN), September 1998.

[19] ETSI, TS 101 350 V6.0.1, Digital Cellular Telecommunications System (Phase 2+), General Packet Radio Service (GPRS), Overall description of the GPRS Radio Interface, August 1998.

[20] ETSI, ETR 015, Radio Equipment and Systems Digital European Cordless Telecommunications (DECT).

[21] ETSI, GSM 01.04 (ETR 350), Digital Cellular Telecommunication Systems (Phase 2+), Abbreviations and Acronyms.

[22] ETSI, GSM 02.02 (ETS 300 904), Digital Cellular Telecommunication Systems (Phase 2+), Bearer Services (BS) Supported by a GSM Public Land Mobile Network (PLMN).

[23] ETSI, GSM 02.03 (ETS 300 905), Digital Cellular Telecommunication Systems (Phase 2+), Teleservices Supported by a GSM Public Land Mobile Network (PLMN).

[24] ETSI, GSM 02.04 (ETS 300 918), Digital Cellular Telecommunication Systems (Phase 2+), General on Supplementary Services.

[25] ETSI, GSM 02.33, Digital Cellular Telecommunication Systems (Phase 2+), Lawful Interception Stage 1.

[26] ETSI, GSM 02.34, Digital Cellular Telecommunication Systems (Phase 2+), High Speed Circuit Switched Data (HSCSD) Stage 1.

[27] ETSI, GSM 03.34 (TS 101 038), Digital Cellular Telecommunication System (Phase 2+), High Speed Circuit Switched Data (HSCSD) Stage 2 Service Description.

[28] ETSI, OSM 03.45 (ETS 300 931), Digital Cellular Telecommunications System, Technical Realization of Facsimile Group 3 Transparent.

[29] ETSI, GSM 07.01 (ETS 300 913), Digital Cellular Telecommunication Systems (Phase 2+), General on Terminal Adaptation Functions (TAF) for Mobile Stations (MS).

[30] Evrett D.B., *Introduction to Smart Card Technology.* Smart Card News, 1999.

[31] Feldmann M., Rissen J.P., *GSM Network Systems and Overall System Integration*. Electrical Communication, 1993.

[32] Galand C., Rosso M., Elie P., *MPE-LTP Coder for Mobile Radio Application.* Speech Communication, vol. 7, 1988.

[33] Ghillebaert B., Maloberti A., Combescure P., *Le système cellulaire numérique européen de communication avec les mobiles*. L'écho des recherches, no. 131, 1988.

[34] Gilhousen K.S., *On the Capacity of a Cellular CDMA System*. IEEE Transaction on Vehicular Technology, 40(2) May 1991, pp. 303–312.

[35] Girault M., Campana M., Bauval A., *Cryptographie et carte à mémoire.* L'écho des recherches, no. 124, 1988.

[36] Greenstein J., Gitlin R.D., *A Microcell/Macrocell Cellular Architecture For Low and High Mobility Wireless Users*. IEEE Global Telecommunication Conference (GLOBECOM 91), IEEE Publishing Company, Conference Record cat. No. 91CH2980-1, vol. 2, pp. 1006–1011.

[37] Greenstein L.J., *Microcells in Personal Communications Systems*. IEEE Communications Magazine, IEEE, New York, December 1992, pp. 76–88.

[38] Harris I., *Data in the GSM Cellular Network*. Editions Cellular Radio Systems, Artech House, Boston, MA, 1993.

[39] Haug T., *Overview of the GSM Project*. EUROCON 88, June 1988.

[40] Huber J.F., *Advanced Equipment for an Advanced Network*. Telcom Report International, 15(3–4) 1992.

[41] Jagoda, A., Villepin, M. de, *Mobile Communications*. John Wiley & Sons, Chichester, 1993.

[42] Jan A., Audestad. *Network Aspects of the GSM System*. EUROCON 88, June 1988.

[43] Jolie P., Mazziotto G., *Une application de la carte à microprocesseur: le module d'identité d'abonné du radiotéléphone numérique européen*. L'écho des recherches, no. 139, 1990.

[44] Lagrange X., Godlewsky P., Tabbane S., *Réseaux GSM-DCS*. Hermes.

[45] Lee W.C.Y., *Overview of the Cellular CDMA*. IEEE Transaction on Vehicular Technology, 40(2) May 1991, pp. 291–302.

[46] Lobensommer H., Mahner H., *GSM – A European Mobile Radio Standard for the World Market*. Telcom Report International, 15(3–4) 1992.

[47] Mac Donald H.V., *The Cellular Concept*. Bell System Technical Journal, 58(1) January 1979.

[48] Mallinder B.J.T., *Specification Methodology Applied to the GSM System*. EUROCON 88, June 1988.

[49] Meyer C., Matyas S., *Cryptography: a New Dimension in Computer Data Security*. John Wiley & Sons, New York.

[50] MOBILENNIUM, *The UMTS Forum Newsletter* (www.umtsforum.org).

[51] Mouly M., Pautet M.B., *The GSM Protocol Architecture, Radio Subsystem Signaling*. IEEE VTC Conference, Saint-Louis, MO, May 1991.

[52] Mouly M., Pautet M.B., *The OSM System for Mobile Communications*.

[53] Natvig J.E., Stein H., De Brito J., *Speech Processing in the Pan-European Digital Mobile Radio System (GSM) – System Overview*. IEEE GLOBECOM 89, November 1989.

[54] Nilsson T., *Toward a New Era in Mobile Communication*. Ericsson WWW Server (http://193.78.100.33/).

[55] NOKIA, *General Packet Radio Service GPRS*. White Paper (www.nokia.com).

[56] NOKIA, *Wireless Application Protocol, the Corporate Perspective*. White Paper (www.nokia.com).

[57] NOKIA, *Wireless Messaging Solutions*. White Paper (www.nokia.com).

[58] Picibono B., *Eléments de théorie du signal*. Dunod, 1977.

[59] *Proceeding of the Digital Cellular Radio-Conference*, Hagen, Westphalie, 12–14 October 1988.

[60] Radix J.C., *Introduction au filtrage numérique*. Eyrolles, 1970.

[61] Rahnema M., *Overview of the GSM System and Protocol Architecture*. IEEE Communications Magazine, April 1993.

[62] Remy J.G., Cueugniet J., Siben C., *Système de radiocommunications avec les mobiles*. Eyrolles.

[63] Rivest R., Shamir A., Aidleman L., *A Method for Obtaining Digital Signatures and Public Key Cryptosystems*. Communication of the ACM, February 1978, pp. 120–126.

[64] Ryan P., *Pan-European Cellular Radio Network*. FITCE, no. 1, 1988.

[65] Schmid E.H., Kähler M., *GSM Operation and Maintenance*. Electrical Communication, 1993.

[66] Seshadri M., Ravi J., *Two User Location Strategies for Personal Communication Services*. IEEE Personal Communications, 1(1), 1994.

[67] Shannon C.E., *A Mathematical Theory of Communication*, Bell System Technical Journal, no. 27, 1948, pp. 379–423 and 623–656.

[68] Silventoinen M., *Personal e-mail, Quoted from European Mobile*. Communications Business and Technology Report, March and December 1995.

[69] Southcott C.B. et al., *Voice Control of the Pan-European Digital Mobile Radio System*. IEEE GLOBECOM 89, November 1989.

[70] Vary P., *Speech Codec for the European Mobile Radio System*. IEEE GLOBECOM 89, November 1989.

[71] Vary P., *A Regular Pulse-Excited Linear Code*. Speech Communication, vol. 7, 1988.

[72] Viterbi A.J., Omura J.K., *Principes des communications numériques*. Dunod.

[73] Watson C., *Radio Equipment for GSM*. Editions Cellular Radio Systems, Artech House, Boston, MA, 1993.

[74] Winch R.G., *Telecommunication Transmission Systems*. McGraw-Hill, New York, 1993.

[75] Witgall M., *The Pan-European Digital Mobile Communication System International Telecom Symposium*, ITS'88, Taipei, Taiwan, 1988.

[76] Young W.R., *Advanced Mobile Phone Service: Introduction, Background and Objectives*. Bell System Technical Journal, 58(1) January 1979, pp. 1–14.

Glossary

ACA	Automatic Channel Assignment
ACELP	Algebraic Code Excited Linear Predictive
ACTS	Advanced Communications Technologies and Services
ADPCM	Adaptive Differential Pulse Code Modulation
AGCH	Access Grant Channel
AMPS	Advanced Mobile Phone System (800 MHz)
API	Application Programming Interface
ARQ	Automatic Repeat Request
ASIC	Application Specific Integrated Circuit
ASPeCT	Advanced Security for Personal Communication Technologies
AUC	Authentication Centre (also AuC)
BCCH	Broadcast Control Channel
BSC	Base Station Controller
BSS	Base Station (Sub)system
BTS	Base Transceiver Station
CAI	Common Air Interface
CDMA	Code Division Multiple Access
CDPD	Cellular Digital Packet Data
CELP	Code Excitation Linear Prediction
CEPT	Conference Européen des Postes et Télécommunications
CGI	Common Gateway Interface
CMIP	Communication Management Information Protocol
CMIS	Communication Management Information Service
CN	Core Network
Codec	Coder/Decoder
CRC	Cyclic Redundancy Check

CSD	Circuit-Switched Cellular Data
CTIA	Cellular Telecommunications Industry Association
CT(*n*)	Cordless Telephony (*n*th generation)
CUG	Closed User Group
DCA	Dynamic Channel Assignment
DCCH	Dedicated Control Channel
DCE	Data Communications Equipment
DCS	Digital Cellular System
DECT	Digital Enhanced Cordless Telecommunications
	Digital European Cordless Telephone
DES	Data Encryption Standard
DNS	Distributed Name Service
DQPSK	Differential QuadraPhase Shift Keying
DSP	Digital Signal Processing
DTE	Data Terminal Equipment
ECMA	European Computer Manufacturers Association
EDGE	Enhanced Data rates for Global Evolution
EEPROM	Electrically Erasable Programmable Read Only Memory
EIR	Equipment Identity Register
EPROM	Erasable Programmable Read Only memory
ERMES	European Radio MEssaging System
ETSI	European Telecommunications Standards Institute
FACCH	Fast Associated Control Channel
FCA	Fixed Channel Assignment
FCCH	Frequency Control Channel
FDD	Frequency Division Duplex
FDMA	Frequency Division Multiple Access
FM	Frequency Modulation
FPLMTS	Future Public Land Mobile Telecommunications System
FTAM	File Transfer Access Method
GGSN	Gateway GPRS Support Node
GMSC	Gateway Mobile Switching Centre
GMSK	Gaussian filtered Minimum Shift Keying
GPRS	General Packet Radio Service
GPS	Global Positioning System
GSM	Global System for Mobile communications
HDLC	High-level Data Link Control
HDML	Handheld Markup Language
HLR	Home Location Register
HSCSD	High Speed Circuit Switched Data
HTML	HyperText Mark-up Language
HTTP	HyperText Transfer Protocol (RFC2068)
IANA	Internet Assigned Number Authority
IDEN	Integrated Digital Enhanced Network
IETF	Internet Engineering Taskforce

IMC	Internet Mail Consortium
IMEI	International Mobile Equipment Identity
IMSI	International Mobile Subscriber Identity
IMT-2000	International Mobile Telecommunication System 2000
IP	Internet Protocol
ISDN	Integrated Switched Digital Network
ISO	International Standards Organisation
ISP	Internet Service Provider
ITU	International Telecommunications Union
IWF	InterWorking Function
LAI	Location Area Identity
LAP	Link Access Protocol
LD-CELP	Low Delay Code Excited Linear Prediction
LEO	Low Earth Orbit
LLC	Logical Link Control
MAC	Media Access Control
MAHO	Mobile Assisted HandOver
MAP	Mobile Application Part
ME	Mobile Equipment
MExE	Mobile Execution Environment
MIB	Management Information Database
MMI	Man-Machine Interface
MS	Mobile Station
MSC	Mobile Switching Centre
MSG(msg)	MeSsaGe
MT	Mobile Terminal
MTP	Message Transfer Protocol
NEP	Noise-Equivalent Power
NMC	Network Management Centre
NMT	Nordic Mobile Telephone System
NSS	Network Subsystem
OMC	Operations and Maintenance Centre
OSI	Open Systems Interconnect
PCH	Paging Channel
PCM	Pulse Code Modulation
PCN	Personal Communication Network
PCS	Personal Communications System
PDA	Personal Digital Assistant
PDC	Program Delivery Control
PDIF	Product Definition Interchange Format
PDN	Packet Data Network
PDU	Protocol Data Unit
PHS	Personal Handyphone System
PIN	Personal Identification Number
PLMN	Public Land Mobile Network

PMR	Private Mobile Radio
PROM	Programmable Read Only Memory
PSK	Phase Shift Keying
PSTN	Public Switched Telephone Network
QAM	Quadrature Amplitude Modulation
RACH	Random Access Channel
RAM	Random Access Memory
RAN	Radio Access Node
RFC	Request For Comments
RLC	Radio Link Control
ROM	Read Only Memory
RPE-LTP	Regular Pulse Excited-Long Term Prediction
SACCH	Slow Associated Control Channel
SAP	Service Access Point
SCH	Synchronisation Channel
SDCCH	Standalone Dedicated Control Channel
SGML	Standardised Generalised Markup Language (ISO8879)
SGSN	Serving GPRS Support Node
SIM	Subscriber Identification Module
SIMEG	Subscriber Identity Module Expert Group
SMC	Short Message Centre
SMGC	Short Message Generating Centre
SMS	Short Message System
SMSC	Short Message Service Centre
SNDC	SubNetwork Dependent Convergence
SNDCP	SubNetwork Dependent Convergence Protocol
SSL	Secure Socket Layer
TACS	Total Access Cellular System
TCH	Traffic Channel
TCP	Transmission Control Protocol
TCP-IP	Transmission Control Protocol-Internet Protocol
TDD	Time Division Duplex
TDMA	Time Division Multiple Access
TE	Terminal Equipment
TeleVAS	Telephony Value Added Services
TETRA	Terrestrial Trunked Radio
TIA	Telecommunications Industry Association
TLS	Transport Layer Security
TLLI	Temporary Link Level Identity
TMN	Telecommunication Management Network
TMSI	Temporary Mobile Subscriber Identification
TSUNAMI	Technology in Smart Antennas for Universal Advanced Mobile Infrastructure
UDP	User Datagram Protocol
UDP-IP	User Datagram Protocol-Internet Protocol

UIM	User Identity Module
UMTS	Universal Mobile Telecommunication System
UPT	Universal Personal Telecommunications
URI	Uniform Resource Identifier (RFC2396)
URL	Uniform Resource Locator (RFC2396)
URN	Uniform Resource Name
USECA	UMTS SECurity Architectures
USSD	Unstructured Supplementary Service Data
UTRA	UMTS Terrestrial Radio Access
UWCC	Universal Wireless Communications Consortium
VLR	Visitor Location Register
VPN	Virtual Private Network
VSELP	Vector Excited Linear Predictor
W3C	World Wide Web Consortium
WAE	Wireless Application Environment
WAP	Wireless Application Protocol
WARC	World Administration of Radio Conference
WBMP	Wireless BitMaP
W-CDMA	Wideband Code Division Multiple Access
WDP	WAP Datagram Protocol
WML	Wireless Markup Language
WSP	Wireless Session Protocol
WTA	Wireless Telephony Applications
WTAI	Wireless Telephony Applications Interface
WTLS	Wireless Transport Layer Security
WTP	Wireless Transaction Protocol
WWW	World Wide Web
XML	Extensible Markup Language

Index

RECEIVED
JUL 2 0 2001